T0139195

# MECHANICAL ANALYSIS OF ELECTRONIC PACKAGING SYSTEMS

# MECHANICAL ENGINEERING

A Series of Textbooks and Reference Books

*Editor*

## L. L. Faulkner

*Columbus Division, Battelle Memorial Institute
and Department of Mechanical Engineering
The Ohio State University
Columbus, Ohio*

1. *Spring Designer's Handbook*, Harold Carlson
2. *Computer-Aided Graphics and Design*, Daniel L. Ryan
3. *Lubrication Fundamentals*, J. George Wills
4. *Solar Engineering for Domestic Buildings*, William A. Himmelman
5. *Applied Engineering Mechanics: Statics and Dynamics*, G. Boothroyd and C. Poli
6. *Centrifugal Pump Clinic*, Igor J. Karassik
7. *Computer-Aided Kinetics for Machine Design*, Daniel L. Ryan
8. *Plastics Products Design Handbook, Part A: Materials and Components; Part B: Processes and Design for Processes*, edited by Edward Miller
9. *Turbomachinery: Basic Theory and Applications*, Earl Logan, Jr.
10. *Vibrations of Shells and Plates*, Werner Soedel
11. *Flat and Corrugated Diaphragm Design Handbook*, Mario Di Giovanni
12. *Practical Stress Analysis in Engineering Design*, Alexander Blake
13. *An Introduction to the Design and Behavior of Bolted Joints*, John H. Bickford
14. *Optimal Engineering Design: Principles and Applications*, James N. Siddall
15. *Spring Manufacturing Handbook*, Harold Carlson
16. *Industrial Noise Control: Fundamentals and Applications*, edited by Lewis H. Bell
17. *Gears and Their Vibration: A Basic Approach to Understanding Gear Noise*, J. Derek Smith
18. *Chains for Power Transmission and Material Handling: Design and Applications Handbook*, American Chain Association
19. *Corrosion and Corrosion Protection Handbook*, edited by Philip A. Schweitzer
20. *Gear Drive Systems: Design and Application*, Peter Lynwander
21. *Controlling In-Plant Airborne Contaminants: Systems Design and Calculations*, John D. Constance
22. *CAD/CAM Systems Planning and Implementation*, Charles S. Knox
23. *Probabilistic Engineering Design: Principles and Applications*, James N. Siddall
24. *Traction Drives: Selection and Application*, Frederick W. Heilich III and Eugene E. Shube
25. *Finite Element Methods: An Introduction*, Ronald L. Huston and Chris E. Passerello

## Additional Volumes in Preparation

*Handbook of Hydraulic Fluid Technology*, edited by George E. Totten

*Thermodynamics*, Earl L. Logan

*Mechanical Engineering Software*

*Spring Design with an IBM PC*, Al Dietrich

*Mechanical Design Failure Analysis: With Failure Analysis System Software for the IBM PC*, David G. Ullman

# MECHANICAL ANALYSIS OF ELECTRONIC PACKAGING SYSTEMS

## STEPHEN A. McKEOWN

*Lockheed Martin Control Systems*
*Johnson City, New York*

CRC Press
Taylor & Francis Group
Boca Raton  London  New York

CRC Press is an imprint of the
Taylor & Francis Group, an **informa** business

CRC Press
Taylor & Francis Group
6000 Broken Sound Parkway NW, Suite 300
Boca Raton, FL 33487-2742

First issued in paperback 2019

© 1999 by Taylor & Francis Group, LLC
CRC Press is an imprint of Taylor & Francis Group, an Informa business

No claim to original U.S. Government works

ISBN-13: 978-0-8247-7033-4 (hbk)
ISBN-13: 978-0-367-39981-8 (pbk)

This book contains information obtained from authentic and highly regarded sources. Reasonable efforts have been made to publish reliable data and information, but the author and publisher cannot assume responsibility for the validity of all materials or the consequences of their use. The authors and publishers have attempted to trace the copyright holders of all material reproduced in this publication and apologize to copyright holders if permission to publish in this form has not been obtained. If any copyright material has not been acknowledged please write and let us know so we may rectify in any future reprint.

Except as permitted under U.S. Copyright Law, no part of this book may be reprinted, reproduced, transmitted, or utilized in any form by any electronic, mechanical, or other means, now known or hereafter invented, including photocopying, microfilming, and recording, or in any information storage or retrieval system, without written permission from the publishers.

For permission to photocopy or use material electronically from this work, please access www.copyright.com (http://www.copyright.com/) or contact the Copyright Clearance Center, Inc. (CCC), 222 Rosewood Drive, Danvers, MA 01923, 978-750-8400. CCC is a not-for-profit organization that provides licenses and registration for a variety of users. For organizations that have been granted a photocopy license by the CCC, a separate system of payment has been arranged.

**Trademark Notice:** Product or corporate names may be trademarks or registered trademarks, and are used only for identification and explanation without intent to infringe.

**Visit the Taylor & Francis Web site at**
**http://www.taylorandfrancis.com**

**and the CRC Press Web site at**
**http://www.crcpress.com**

*To*
*my wife, my children,*
*and my Lord Jesus Christ,*
*who made this possible*

# Preface

One of the problems I have frequently encountered in over 20 years of mechanical engineering analysis is the tendency to make a simple situation more complex by using a sophisticated computer tool (such as finite-element analysis) without first considering the real situation. In some cases, an analysis has been delayed due to the complexity and/or limitations of the software much longer than necessary, resulting in excessive costs or schedule delays. In other situations, the computer tool was misapplied resulting in incorrect results. It is my intent to demonstrate how basic engineering can be applied to engineering analysis using a combination of time-honored techniques in conjunction with modern computing power to achieve cost-effective results in a timely manner.

This book is intended as a practical guide for engineers working in the field of electronic packaging. Emphasis is given to the ability of electronic packaging configurations to withstand the thermal, mechanical, and life environments encountered in everyday use. Although it includes analytical techniques for many of the design needs for electronic packaging, it is hoped that an efficient way of thinking before acting can be conveyed. Often, a complex analytical technique is used when a much simpler approach can give equal or better results with much less time and cost. It is also hoped that the reader can use this book to develop analytical techniques based upon basic engineering principles. It is important to realize that engineering analysis is not a "plug and chug" approach into a formula or a commercial software package but requires understanding the underlying approach and assumptions. Emphasis is given to the practical use of computers for automating the solution process, bridging the gap between purely empirical and detailed texts explaining the use of software packages. Since this book treats the basic principles and methods of attack, it may benefit those in fields other than electronic packaging. A discussion of how the analytical techniques described can be applied to recent electronic packaging developments such as ball-grid arrays and multichip modules is also included. A brief description of each chapter follows.

Chapter 1 introduces a typical electronic packaging configuration, provides examples of electronic packaging analysis, and describes planning the analysis. Included is a definition of the problem and determination of the analytical method along with required input and output data, and cost and accuracy considerations.

Chapter 2 describes the tools used for engineering analysis and compares the advantages and disadvantages of each tool. Included are hand calculations, symbolic equation solvers, spreadsheets, finite-element analysis, and custom programs. A discussion of "seamless" automation techniques is also included.

Chapter 3 describes methods for determining the thermal performance of electronic equipment at each level of packaging (component, board/module, and chassis) using both classical and finite-element techniques. Also included is a discussion of special cooling techniques such as flow-through modules, heat pipes, thermoelectrics, and immersion/phase change cooling.

Chapter 4 discusses methods for predicting the mechanical performance of electronic packaging configurations in various static and dynamic environments. This includes the use of both classical and finite-element methods, and special cases such as thermal shock or acoustic noise.

Chapter 5 explains the techniques for analysis of component life in thermal cycling and vibration. Although solder joint life is emphasized, methods are presented for life analysis of structural members and plated-through holes. Also discussed is the life of component leads under combined thermal cycling and vibration, and the determination of expected life by comparison to test data.

Chapter 6 describes techniques that cannot be categorized as traditional thermal, mechanical, or life analysis. Included in this chapter are fire resistance, humidity, and pressure-drop analysis. The use of similarity and energy-based methods in performing engineering analyses is also discussed.

Chapter 7 covers the setup and analysis of tests to measure or verify system performance. Included in this chapter are discussions of thermal, vibration, life, and expansion data collection and analysis.

Chapter 8 discusses the preparation of reports of engineering analysis and the verification of results. This chapter emphasizes the importance of documenting and verifying source data.

Example problems and solutions are included throughout this book to emphasize the various analytical approaches presented. Although a number of specific situations are analyzed in this book, an understanding of basic engineering principles is required to properly apply the concepts described in this book to specific situations. The checklists at the end of each chapter highlight potential problem areas and help avoid common errors. It must be noted that any engineering analysis is only as good as the input data and underlying assumptions, and all results should be confirmed by testing.

I would like to thank Gary Miller and my wife for their editorial contributions throughout the writing of this book. I would also like to thank Len Stefik, John Lee, Ray Heimbach, Rich Luybli, Leo McDermott, Gary Stefani, Dan Hogan, David Guy, and Paul Checkovich for their support and comments.

# Contents

# MECHANICAL ANALYSIS OF ELECTRONIC PACKAGING SYSTEMS

# Chapter 1
# Introduction

## 1.1   BACKGROUND

Electronic packaging is "the technology of packaging electronic equipment" [1] which includes the interconnection of electronic components into printed wiring boards (PWBs) and the interconnection of printed wiring boards to electronic assemblies. Because of the increased use of computers and electronics in all aspects of our lives, increasing performance of electronic packaging configurations without increasing cost is becoming a major thrust of the electronics industry. This increased use of electronics in conjunction with reduced size and weight, and the business need to introduce products rapidly has led to increased use of analysis instead of "build and test" to determine if equipment can meet the appropriate requirements. Performing this analysis accurately and in a timely manner is very important if the full benefits of an analytical approach are to be realized.

### 1.1.1   Typical Electronic Packaging Configuration

An electronic packaging configuration typically consists of component packages containing silicon die ("chips"), and other components such as capacitors and resistors, mounted on printed wiring boards. These printed wiring boards with components (called modules) are then mounted in a chassis which provides protection from the environment, cooling, mechanical support, and a method of interfacing to the outside world (see Figure 1-1).

### 1.1.2   Electronic Packaging Levels

Because of differences in materials and configuration requiring a different analytical approach, an electronic packaging configuration is frequently broken down into the following levels:

- Component (level 1)
- Module (level 2)
- Chassis (level 3)

Although the above packaging levels are frequently used, they are not universal. Increasing packaging density and the use of novel packaging approaches (such as multichip modules and chip on board) has often clouded the distinction between the packaging levels.

1

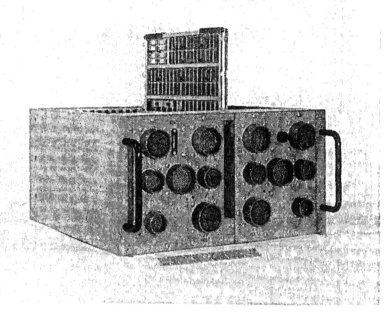

**Figure 1-1 - Typical Electronic Packaging Configuration**
(Courtesy of Lockheed Martin Control Systems.)

**Component (First) Level Packaging**

Component-level packaging provides a method for attaching and interconnecting a silicon microcircuit to the next packaging level, and protecting the microcircuit from the environment. The configuration of the component-level package and materials selection depends upon whether hermetic (ceramic) or non-hermetic (plastic) packaging technology is used.

For a hermetic packaging configuration (see Figure 1-2), the silicon die is bonded into the cavity of a ceramic package and the configuration is sealed with a lid that closely matches the expansion rate of the package. Typically, small wires (bond wires, not shown) are used to interconnect the pads on the silicon die to the leads. In some cases, leads are not used and interconnection to the printed wiring board is made through metallized areas on the outside of the package (leadless chip-carriers).

In a non-hermetic configuration (see Figure 1-2), the silicon die is bonded to a heat spreader which is typically part of a lead frame. Interconnection is made either with small wires between the die pads and the leads or directly between the leads and pads (see 1.1.3). Once interconnection is made, the entire assembly is encapsulated in an epoxy material to provide protection from the environment.

Because ceramic is a brittle material, the package lid and leads must match the expansion of the ceramic. Typical materials used in component-level packaging are summarized in Table 1-1.

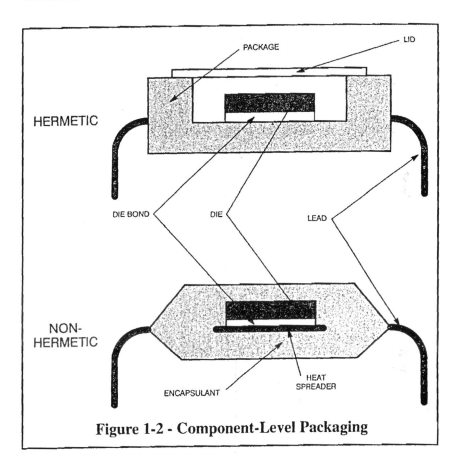

**Figure 1-2 - Component-Level Packaging**

## Module (Second) Level Packaging

Module-level packaging interconnects components to the next level of packaging. Specifics of the module packaging configuration vary if through-hole or surface-mount technology is used.

A through-hole technology module typically uses dual-inline-package (DIP) packaging for the microcircuits (see Figure 1-3). The leads on these components are soldered into holes in a printed wiring board (PWB) to provide interconnection to other components and the chassis (via wires or connectors). In cases where thermal performance needs to be enhanced, a heat sink may be attached to the PWB directly underneath the component.

| Table 1-1 - Typical Component-Level Packaging Materials | | |
|---|---|---|
| | **Hermetic** | **Non-Hermetic** |
| **Package Body** | Aluminum oxide (alumina) | Epoxy |
| **Package Lid** | Nickel-iron alloy* | - |
| **Leads** | Nickel-iron alloy* | Copper |
| **Heat spreader** | - | Copper |
| **Die bond** | Solder or epoxy | Epoxy |
| **Die** | Silicon | Silicon |
| * - Specifically formulated to provide expansion match to ceramic | | |

Surface-mount technology (SMT) solders components directly to pads on the surface of the PWB (see Figure 1-3). Since leads are not inserted into holes in the PWB, components may be mounted on the back surface of the module without concern for the location of parts on the front. Since SMT parts are typically smaller than DIP parts and components may be mounted on both sides of the module, an increase in packaging density by approximately a factor of 3 is realized. Thermal performance in a SMT module is typically accomplished by bonding a heat sink between two surface-mount PWBs, or by laminating a heat sink as part of the PWB fabrication process. In some cases additional plated-through holes (PTHs) are used to enhance the thermal conduction between the component and heat sink (see Figure 3-5). Since there is often very little compliance between the component and the module surface in SMT applications, it is often important that the expansion of the module surface match the component. Typical materials used in module level packaging are summarized in Table 1-2.

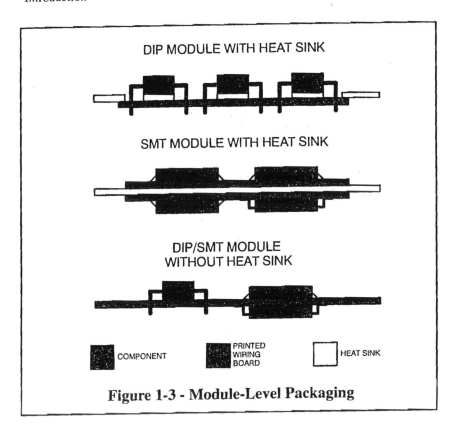

**Figure 1-3 - Module-Level Packaging**

| Table 1-2 - Typical Module-Level Packaging Materials | | |
|---|---|---|
| | **Through-Hole and SMT** | **Hermetic SMT\*** |
| **PWB** | Glass/epoxy<br>Glass/polyimide | Glass/epoxy or glass/polyimide with CIC or CMC planes<br>Aramid/epoxy |
| **Heat Sink** | Aluminum<br>Copper | Copper-invar-copper (CIC)<br>Copper-molybdenum-copper (CMC)<br>Silcon-carbide reinforced aluminum |
| \* - Configurations exposed to thermal cycling | | |

### Chassis (third) Level Packaging

Typically the chassis-level package (see Figure 1-4) includes support rails to which the modules are mounted, and a motherboard with connectors that provides electrical interconnection to other modules and the main chassis connectors (not shown). Modules may be mounted to the chassis by spring-loaded clips, mechanically actuated clamps, bolts, or similar methods (not shown). Chassis cooling may be enhanced by fins (see Figure 1-4), tubes, or cooling plenums. Materials used for chassis-level packaging include aluminum, steel, plastics, and composites depending upon cost and weight considerations.

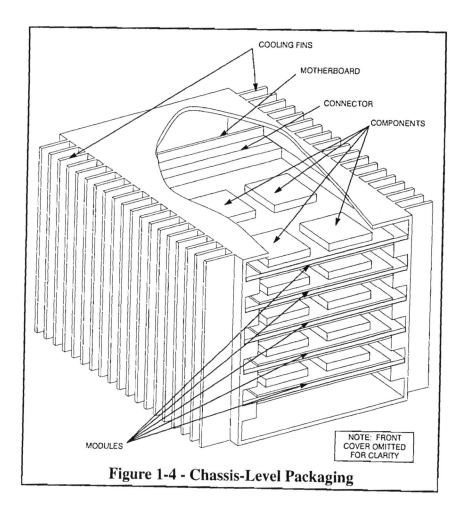

**Figure 1-4 - Chassis-Level Packaging**

## 1.1.3   Recent Development in Electronic Packaging

Although much of the analysis in this book is at the module and chassis level, there have been some electronic packaging configurations developed in the past few years that may require special consideration. Specifically, these configurations include:

- Ball-Grid Arrays (BGA)

- Multichip Modules (MCM)

- Flip-Chips

- Chip-on-Board (COB)

Use of the above configurations may require special considerations in thermal, vibration, and solder life analyses.

**Ball-Grid Arrays [2]**

A ball-grid array (BGA, see Figure 1-5) is characterized by the array of solder balls on the bottom of the package that is used to make interconnection to the PWB. Typically, these solder balls have a higher melting point than the tin-lead eutectic solder used to make the interconnection. This allows the height of the solder joint to be controlled during the soldering process. In some cases, an under-fill may be applied between the part and the PWB to improve thermal perfor-mance, mechanical durability, and/or solder joint life. In other cases, a combina-tion of "dimpling" the underside of the package and close-tolerance solder mask can provide lateral support to the solder which may improve solder life.

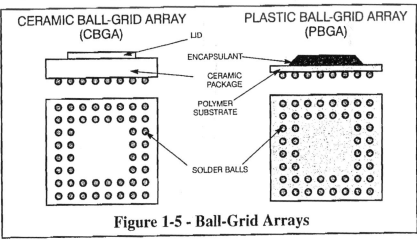

**Figure 1-5 - Ball-Grid Arrays**

A ceramic ball-grid array package is similar to a conventional ceramic package (see Figure 1-2) since the die is mounted in a similar manner, but the ceramic substrate is configured to make the interconnection through an array of pads on the bottom of the package instead of through leads on the perimeter.

A plastic ball-grid array mounts the die on a polymer substrate (similar to a PWB) and an encapsulant is applied over the top of the die to provide protection from the environment. Since the BGA provides interconnection of an area instead of the perimeter, much higher interconnection densities are available (especially for larger packages). Because of this high interconnection density, special consideration may need to be given to the design of the PWB (such as the use of microvia technology).

**Multichip Modules [2,3]**

A multichip module (MCM, see Figure 1-6) may be simply defined as a package containing more than one chip. By incorporating more than one chip in a single package, the additional packaging area required to interconnect to the PWB is effectively spread over more than one chip and the packaging density is increased. Multichip modules may be grouped into three categories MCM-C, MCM-D, and MCM-L. MCM-C utilizes ceramic or glass-ceramic substrates and fireable materials to provide the interconnection. MCM-D uses deposited metals and unreinforced dielectrics to provide the interconnection with a ceramic, metal, or other rigid base. One such MCM-D configuration, developed by General Electric and Lockheed Martin, incorporates silicon die into cavities in a substrate with interconnection provided by a polyimide/copper system with laser-drilled vias (see Figure 1-6). MCM-L uses reinforced laminates to provide the interconnection (similar to PWB).

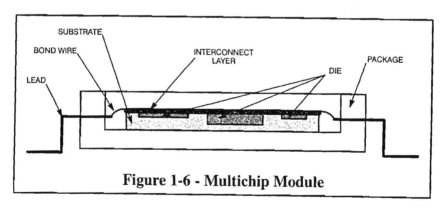

**Figure 1-6 - Multichip Module**

**Flip-Chips [4]**

In a flip-chip configuration (see Figure 1-7), the component die is mounted with the active surface facing the substrate using beam leads or solder bumps. The conventional configuration bonds the rear surface of the die to the substrate and utilizes bond wires to make the interconnection. In some situations (such as plastic encapsulated parts), the beam leads can be part of the lead frame, eliminating the need for the substrate. Since interconnection lengths are shorter in flip-chip configurations, electrical performance is improved and packaging density is increased. Flip-chip technology may or may not be included in BGAs, MCMs, or chip-on-board configurations.

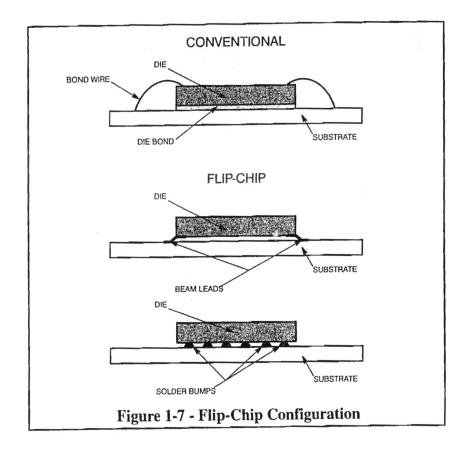

**Figure 1-7 - Flip-Chip Configuration**

**Chip-on-Board [4]**

Chip-on-board (COB) technology mounts the silicon die directly to the PWB (see Figure 1-8) using either a die attach adhesive and wire bonding or flip-chip technology. An encapsulant (sometimes referred to as a "glob top") is applied over the chip and interconnection to provide protection. Because the conventional package is eliminated, packaging density is increased and cost is reduced although repairability is sacrificed.

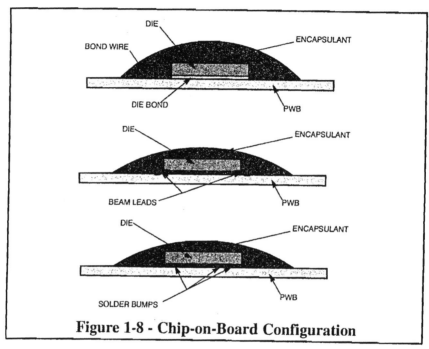

**Figure 1-8 - Chip-on-Board Configuration**

## 1.1.4   Examples of Electronic Packaging Analysis

Electronic packaging analysis may be used to determine the performance of the appropriate hardware at various packaging levels, depending upon the specific analysis to be performed (see Table 1-3). Some of the more common applications of electronic packaging analysis include:

- Thermal performance

- Module vibration performance

- Solder life

| Table 1-3 - Analyses Typically Applied to Various Packaging Levels | | | | |
|---|---|---|---|---|
| | Packaging Level | | | |
| Type of Analysis | Compo-nent | Module | Chassis | Reference |
| Thermal Performance | ● | ● | ● | Chapter 3 |
| Vibration | ● | ● | ● | 4.5.1 − 4.5.3 |
| Acoustic Noise | | | ● | 4.5.4 |
| Mechanical Shock | ● | ● | ● | 4.5.5 |
| Sustained Acceleration | ● | ● | ● | 4.6.2 |
| Thermal Stress | ● | ● | ● | 4.6.3 |
| Solder Life | | ● | | 5.2 |
| Mechanical Hardware Life | ● | ● | ● | 5.3.1 |
| PTH Life | | ● | | 5.3.2 |
| Fire Resistance | | | ● | 6.2 |
| Transducer Rupture | | | ● | 6.3 |
| Humidity | ● | ● | ● | 6.4 |
| Pressure Drop | | | ● | 6.5 |
| Drop Test | ● | ● | ● | 6.7 |

**Thermal Performance Analysis**

An aluminum electronic chassis (see Figure 1-9) contains five surface-mount technology modules. Each module dissipates 25 Watts for a total of 125 Watts. PWBs in the modules are bonded to the aluminum heat sink with 10 mils of silicone rubber. Using thermal performance analysis (see 3.7.4), the steady-state temperatures for operation in a 25°C ambient are determined as follows:

- Average chassis temperature = 56.2°C

- Average module temperature = 60.6°C

If a hypothetical (not shown) surface-mount technology component (0.95 in x 0.95 in) dissipating 1 Watt (with a junction-to-case rise of 5°C/W) is mounted on the module, the junction temperature is expected to be 76.6°C.

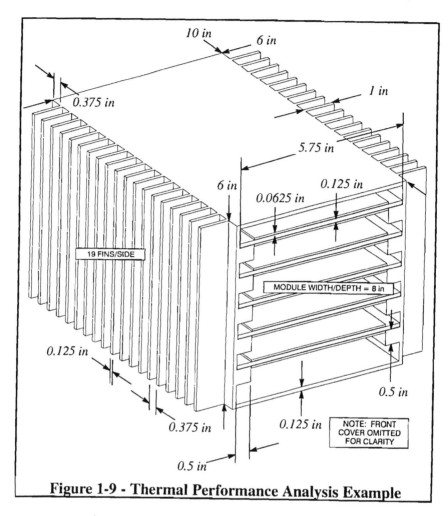

**Figure 1-9 - Thermal Performance Analysis Example**

## Acoustic Performance

A flat aluminum panel 10.5 in x 6.39 in x 0.0625 in thick is part of an electronic chassis. This panel is exposed to an acoustic input (pressure band level) of 140 db from 100 Hz to 2,000 Hz. Using acoustic performance analysis (see 4.7.4) with a damping of 1.5%, the following is determined:

- Overall input sound pressure level = 151.1 db

- Maximum stress in panel = 2,703 psi-RMS

- Maximum deflection of panel = 0.0334 in-RMS

- Maximum acoustic intensity inside chassis = 146.9 db

## Module Vibration Performance

A surface-mount module with glass/epoxy PWBs, an aluminum heat sink, and a silicone rubber bond (see Figure 1-10) is simply supported on three edges and exposed to a random vibration input of 0.04 $g^2$/Hz from 50 Hz to 1,000 Hz. Using vibration performance analysis (see 4.7.3) with a total module mass of 1.16 lbm, and an overall damping ratio of 10%, the following is determined:

- Module primary resonant frequency = 107 Hz

- Maximum module deflection = 0.009 in-RMS

- Maximum PWB surface strain = 54.2 ppm-RMS

## Solder Vibration Life

The surface-mount technology module described above (see Figure 1-10) has a 28-terminal leadless chip-carrier (0.45 in x 0.45 in) mounted in the center. Tin-lead eutectic solder joints are of typical geometry with a height of 0.003 in under the package (50-mil pitch). Using the above vibration environment and a cumulative failure probability of 1%, the solder vibration life is determined to be 4.8 minutes (see 5.6.2).

# 1.2   PLANNING THE ANALYSIS

Prior to any analysis, it is important to establish a plan to ensure an efficient approach and avoid unnecessary or counterproductive effort. Specific steps to be taken include:

- Defining the problem and/or objective of the analysis

- Determining the approach to be used

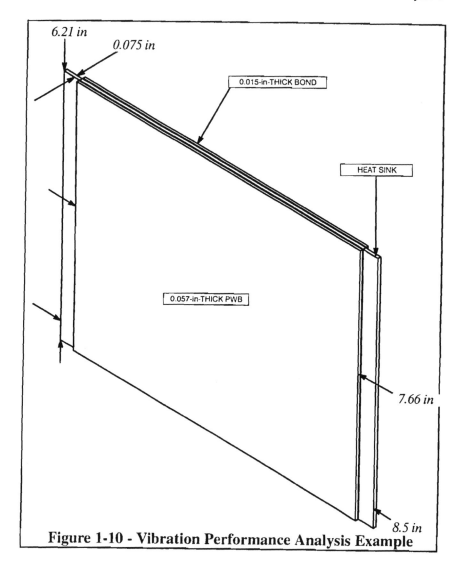

**Figure 1-10 - Vibration Performance Analysis Example**

## 1.2.1    Problem Definition/Objective

Although it may seem trivial, one of the greatest challenges in any engineering analysis is defining what the real problem is. In some cases the problem may seem obvious but pre conceived notions and external pressures may have clouded one's thinking. It is very important to objectively look at the situation to best determine the appropriate problem to solve, and avoid solving a problem that may not be relevant. Once the problem is defined, determining a related objective can help to keep the analysis focused on the needs, and help avoid activities that are not necessary.

## 1.2.2    Analysis Approach

Once the objective of the analysis is defined, an approach for accomplishing the objective can be determined. This includes determining the analytical method to be used and investigating the various cost and accuracy trade-offs.

**Analytical Method**

Planning the method of the analysis includes selecting the type of analysis, and determining the relevant physical principle or field of study required. Included in the plan is the selection of the appropriate tool required to perform the calculations. Typical analytical tools include (see Chapter 2):

- Hand calculations
- Symbolic equation solvers
- Spreadsheets
- Finite-element analysis
- Custom programs

It is important that the appropriate input data is available and the output data is consistent with the problem/objective. Table 1-4 summarizes the input/output data relevant to various analyses.

Although a variety of analytical techniques are available, many of these approaches include inherent assumptions and simplifications. Alternatively, the environment and/or equipment may be too complex to analyze within the available resources. In these situations, testing may be required to obtain the necessary data or verify the analytical results (see Chapter 7).

**Cost and Accuracy Considerations**

Background

Selection of an approach for conducting an analysis depends upon a trade-off of the relative cost and accuracy of the various analytical methods. Usually, increasing the accuracy of an analysis increases the cost of the analysis. This increased cost can be from a combination of the human and computer resources required to complete the task.

## Analytical Method

Typically the cost/accuracy trade-off affects the selection of the method used to complete the analysis. In some cases, a quick analysis using a classical analytical technique and a symbolic equation solver or spreadsheet will show that there is a large margin between the expected performance and requirements. In other cases where the performance/requirement margin is not so large, a more detailed analysis may be required.

## Finite-Element Analysis

Usually finite-element analysis represents the most accurate approach for conducting an analysis since fewer geometric simplifications (usually used for a classical analysis) are required. However, cost/accuracy trade-offs also exist in the application of the finite-element approach. Typically, increasing the density of a finite-element mesh increases the accuracy but also increases the cost required to complete the analysis.

Selection of element type also influences the cost/accuracy trade-off. A finite-element mesh using tetrahedral elements can be easily generated for a solid using automatic meshing available from various finite-element pre-processors and CAD software, but the results obtained may not be as accurate as those using brick elements. Unfortunately, generation of finite-element mesh using brick elements is very labor-intensive for a complex geometry.

## Effect of Tolerances

Even if the expected accuracy of an analytical method is sufficient, it is possible that variations due to tolerances may be significant. This is especially true when the tolerance is a large fraction of the dimension. In cases where tolerances are significant or the analysis shows a small performance/requirement margin, the analysis should be repeated for the best and worst-case combinations of dimensions.

## Table 1-4 - Typical Input and Output Data

| Type of Analysis | Input Data | Output Data |
|---|---|---|
| Thermal Performance (steady-state) | • Boundary temperatures<br>• Heat loads<br>• Fluid flow rates/velocities<br>• Thermal conductivity<br>• Fluid density<br>• Fluid specific heat<br>• Viscosity of cooling fluid<br>• Expansion of cooling fluid<br>• Surface emissivity<br>• Geometry | • Chassis temperature<br>• Module temperatures<br>• Component temperatures |
| Thermal Performance (transient) | • Boundary temperatures<br>• Heat loads<br>• Duration of heat loads/temperatures<br>• Fluid flow rates/velocities<br>• Thermal conductivity<br>• Density<br>• Specific heat<br>• Viscosity of cooling fluid<br>• Expansion of cooling fluid<br>• Surface emissivity<br>• Geometry | • Chassis temperature vs. time<br>• Module temperatures vs. time<br>• Component temperatures vs. time |
| Vibration | • Vibration amplitude vs. frequency<br>• Vibration duration<br>• Vibration direction<br>• Modulus of elasticity<br>• Poisson's ratio<br>• Density<br>• Damping ratio<br>• Geometry | • Resonant frequency<br>• Response acceleration<br>• Response deflection<br>• Stresses<br>• Strains<br>• Curvatures |

| Table 1-4 - Typical Input and Output Data (continued) | | |
|---|---|---|
| **Type of Analysis** | **Input Data** | **Output Data** |
| Acoustic Noise | • Acoustic amplitude<br>• Duration of acoustic amplitude<br>• Altitude<br>• Modulus of elasticity<br>• Poisson's ratio<br>• Density<br>• Damping ratio<br>• Speed of sound<br>• Geometry | • Response deflection<br>• Stresses<br>• Transmitted acoustic intensity |
| Mechanical Shock | • Shock amplitude vs. time<br>• Number of shocks<br>• Shock direction<br>• Modulus of elasticity<br>• Poisson's ratio<br>• Density<br>• Damping ratio<br>• Geometry | • Response deflection<br>• Stresses |
| Sustained Acceleration | • Acceleration amplitude<br>• Acceleration direction<br>• Modulus of elasticity<br>• Poisson's ratio<br>• Density<br>• Geometry | • Response deflection<br>• Stresses |
| Thermal Stress | • Initial/final temperature<br>• Temperature ramp rate<br>• Modulus of elasticity<br>• Poisson's ratio<br>• Density<br>• Expansion rate<br>• Thermal conductivity<br>• Specific heat<br>• Geometry | • Expansion rate<br>• Response deflection<br>• Stresses |

| Table 1-4 - Typical Input and Output Data (continued) | | |
|---|---|---|
| **Type of Analysis** | **Input Data** | **Output Data** |
| Solder Vibration Life | • PWB surface strain<br>• PWB curvature<br>• Vibration duration<br>• Shear modulus<br>• Vibration life multiplier<br>• Vibration life exponent<br>• Weibull parameters<br>• Geometry | • Expected life<br>• Cumulative failure probability |
| Solder Thermal Cycling Life | • Thermal cycling range<br>• Component temperature rise<br>• Average solder temperature<br>• Thermal dwell time<br>• Number of cycles applied<br>• Expansion rate<br>• Fatigue ductility coefficient<br>• Fatigue ductility exponent<br>• Life correction cycles<br>• Life correction exponent<br>• Geometry | • Cycles to fail<br>• Cumulative failure probability |
| Mechanical Hardware Life | • Vibration stresses<br>• Vibration amplitude<br>• Fatigue multiplier<br>• Fatigue exponent | • Expected life<br>• Cumulative damage |
| PTH Life | • Thermal cycling range<br>• Number of cycles applied<br>• Expansion rate<br>• Modulus of elasticity<br>• Modulus of plasticity<br>• Yield stress<br>• Ductility<br>• Ultimate strength<br>• Geometry | • Cycles to fail<br>• Cumulative damage |

## Table 1-4 - Typical Input and Output Data (continued)

| Type of Analysis | Input Data | Output Data |
|---|---|---|
| Fire Resistance | • Heat input to calibration tube<br>• Temperature of gas<br>• Surfaces of flame impingement<br>• Thermal conductivity<br>• Density<br>• Specific heat<br>• Viscosity<br>• Geometry | • Equipment temperatures<br>• Heat input from flame |
| Transducer Rupture | • Pressure of gas<br>• Ambient pressure<br>• Chassis temperature<br>• Temperature of gas<br>• Density<br>• Specific heat<br>• Specific heat ratio<br>• Universal gas constant<br>• Geometry | • Heat input from gas<br>• Pressure response time |
| Humidity | • Pressure<br>• Temperature<br>• Relative humidity<br>• Humidity duration<br>• Activation energy of material in humidity<br>• Humidity exponent of material | • Dew point<br>• Humidity ratio<br>• Relative humidity<br>• Humidity life |
| Pressure Drop | • Fluid mass flow rate/velocity<br>• Fluid temperature<br>• Fluid density<br>• Fluid viscosity<br>• Geometry | • Pressure drop |

| Table 1-4 - Typical Input and Output Data (continued) | | |
|---|---|---|
| **Type of Analysis** | **Input Data** | **Output Data** |
| Drop Test | • Initial velocity<br>• Change in height<br>• Acceleration due to gravity<br>• Compression strength of wood (if applicable)<br>• Modulus of elasticity<br>• Poisson's ratio<br>• Modulus of plasticity (if applicable)<br>• Yield stress (if applicable) | • Maximum impact force<br>• Maximum impact acceleration<br>• Maximum impact stress |

### Effect of Material Properties and Vendor Data

Some of the key inputs to an analysis include material properties and vendor data (such as thermal resistance values). Any variations in this information may significantly influence results. When such data is used, it is important to understand the specific conditions, environment, or test methods that were used to determine the information. It is also important to note the implied differences between the various data sources. Purchase and material specifications often provide limits for determining the acceptability of material which is helpful in worst-case analyses, whereas vendor data may represent typical or optimistic values. In critical or marginal applications, published data should be verified by testing and/or the analysis should be repeated for the best and worst-case combination of data.

### Effect of Assumptions

All of the assumptions used in the analysis, including underlying assumptions of the approach, should be examined critically to determine if they are consistent with the specific situation.

### Verification

Any analysis plan should include sufficient time for verification to help avoid any inaccuracies that might arise. This verification may be by testing, use of a simplified model, or other methods as appropriate. For information on verification, see Chapter 8.

# 1.3    UNITS

## 1.3.1    Background

In any engineering analysis, it is important to maintain a consistent system of units. Since in many situations input data is obtained with units that are not necessarily consistent, it is recommended that the units be included in any equation solution to verify the consistency of such units. Information on unit conversions may be found in Appendix B.

## 1.3.2    Acceleration Due to Gravity

The acceleration due to gravity correction ($g_c$) used in some texts is not used in this text since it is not required (and may cause confusion) if a consistent system of units is used. This can be illustrated by Newton's first law, as follows:

$$F = ma \qquad (1\text{-}1)$$

$F$ = Force
$m$ = Mass
$a$ = Acceleration

Using a mass of 1 lbm, and an acceleration of 32.2 ft/sec$^2$ the force is calculated as follows:

$$F = ma = 1lbmx32.2\frac{ft}{sec^2} = 32.2\frac{lbm\text{-}ft}{sec^2}$$
$$F = 32.2\frac{lbm\text{-}ft}{sec^2}x\frac{lbf}{32.2\frac{lbm\text{-}ft}{sec^2}} = 1\ lbf \qquad (1\text{-}1)$$

This demonstrates that including the units in the equation and making appropriate unit conversions produces the correct results without explicitly including the $g_c$ factor. The acceleration due to gravity is a consideration in free convection analysis (see equation 3-26 in 3.6.2) and situations involving potential energy (see 6.7.2, 6.7.3, and 6.7.6). Care must be taken to ensure that the proper conversion is used in vibration situations where the acceleration is expressed in terms of the acceleration due to gravity.

# 1.4    CONVENTIONS

Throughout this text, the definition of variables used in the various equations will immediately follow the group of equations. These variable definitions are defined in order of use in the respective equations. A checklist is provided at the end of each chapter that summarizes the information required and potential pitfalls. References are included at the end of the chapter in which they are referred, but are numbered consecutively from the beginning of the book. The citation for any reference used in more than one chapter is repeated. Abbreviations are explained parenthetically at first occurrence and are also summarized in Appendix A.

## 1.5   GENERAL ANALYSIS CHECKLIST

### 1.5.1   Applicable to Analysis Planning

☐ Has the problem or objective of the analysis been properly defined?

☐ Is all the required input data available?

☐ Is the output data consistent with objectives?

☐ Is the expected accuracy of the analytical method sufficient?

☐ Does the analysis plan include consideration of tolerances?

☐ Are the material properties and vendor information consistent with the application?

☐ Are the expected assumptions consistent with the situation?

☐ Has verification been included in the plan?

### 1.5.2   Applicable to Units

☐ Has a consistent system of units been used?

☐ Have the units been included in any equation solution?

☐ Has the proper unit conversion been included for acceleration due to gravity?

☐ Has the proper unit conversion been included for vibration expressed in g-units?

## 1.6   REFERENCES

[1]   Lapedes, D. (ed.); McGraw-Hill Dictionary of Scientific and Technical Terms, Second Edition; McGraw-Hill, 1978

[2]   Lau, J.H. (ed.); Ball Grid Array Technology; McGraw-Hill, 1995

[3]   Messner, Turlik, Balde, Garrou; Thin Film Multichip Modules; International Society for Hybrid Microelectronics, 1992

[4]   Manzione, L.T.; Plastic Packaging of Microelectronic Devices; AT&T Bell Laboratories/Van Nostrand Reinhold, 1990

# Chapter 2
# Analytical Tools

## 2.1 INTRODUCTION

The analysis of electronic packaging, as any engineering analysis, uses a variety of analytical tools to perform the required computations. Typical tools used in the analysis of electronic packaging include:

- Hand calculations

- Symbolic equation solvers

- Spreadsheets

- Finite-element analysis

- Custom programs

The selection of the appropriate tool depends upon the specifics of the situation and the resources available. In many situations, a combination of the above tools are used. A spreadsheet may be used to evaluate the results of a finite-element analysis or an equation developed using a symbolic equation solver. In addition to the technical reasons for selecting an analytical tool, another important consideration is the ability to "seamlessly" transfer data between applications. This is especially true with the increased use of computer data bases for geometry and other information.

## 2.2 ANALYTICAL TOOLS

### 2.2.1 Hand Calculations

**Description**

Hand calculations include the equation derivations, sketches, and calculations typically written by hand on a data pad, back of an envelope, napkin, etc. Although no particular equipment is required, sometimes the numerical calculations can be assisted by the use of an electronic calculator.

**Advantages of Hand Calculations**

- No special tools are required (except a calculator) so an analysis can be conducted equally well in the lunchroom or on top of a mountain as in an office

- Familiarity with a specific software package or computer is not required so there is no "learning curve"

- Concepts can be developed rapidly since it is easy to include sketches, Greek letters, and other special symbols

**Disadvantages of Hand Calculations**

- Once an analysis is completed, it may be difficult to make changes without repeating almost all of the work

- Hand calculations are difficult to check since the work may need to be repeated to verify the results

- Since hand calculations are a manual process, they are subject to a variety of human errors including incorrect copying of equations and mathematical errors

- Unit conversions must be performed and tracked manually

- Processing large volumes of data can be tedious and/or time-consuming

## 2.2.2    Symbolic Equation Solvers

**Description**

A symbolic equation solver is a software product that allows the user to solve algebraic equations, conduct integration and differentiation, etc. symbolically instead of numerically as is typical of computer-based methods.  The symbolic equation solver may be thought of as computer-based hand calculations since equation representations can be made in a manner that is very much like those written and derived by hand.  In addition to performing symbolic solutions, specific values for the variables can be included (along with appropriate units) to determine specific solutions. The ability to evaluate at multiple data points and graph results is also typically included. For cases where an algebraic solution is not possible, a numerical solution for a specific data point may be available.  Further information on symbolic equation solvers may be obtained from the software vendors.

**Typical Application of Symbolic Equation Solver**

Symbolic Equation Solution

Castigliano's theorem [5] showed that for linear systems, the deflection under a load can be determined as follows:

$$\delta_F = \frac{\partial U}{\partial F} \qquad (2\text{-}1)$$

$\delta_F$ = *Deflection under Load F*
$U$ = *Elastic Energy Stored in Structure*
$F$ = *Load F*

For the configuration shown in Figure 2-1, the total bending energy is determined as follows [5]:

$$U = \frac{1}{2EI} \int_0^{a+b} (M(x))^2 dx \qquad (2\text{-}2)$$

$$M(x) = Fx \quad (for\ x < a)$$
$$M(x) = Fa \quad (for\ x > a) \qquad (2\text{-}3)$$

$U$ = *Total Bending Energy*
$E$ = *Modulus of Elasticity*
$I$ = *Moment of Inertia*
$a, b$ = *Dimensions (see Figure 2-1)*
$x$ = *Distance Along Length of Beam (from load)*
$M(x)$ = *Bending Moment at Distance x*
$F$ = *Force*

Using a symbolic equation solver (see Figure 2-2), the deflection under the load is found to be:

$$\delta_F = \frac{1}{3EI} Fa^2(a + 3b) \qquad (2\text{-}4)$$

$\delta_F$ = *Deflection under Load F*
$E$ = *Modulus of Elasticity*
$I$ = *Moment of Inertia*
$F$ = *Force*
$a, b$ = *Dimensions (see Figure 2-1)*

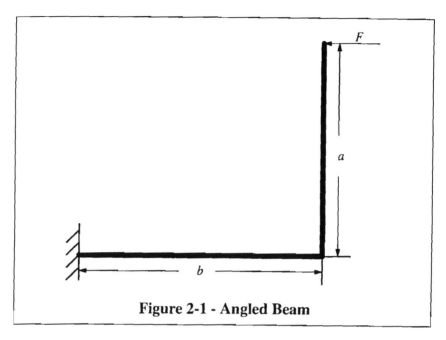

**Figure 2-1 - Angled Beam**

Equation Evaluation (with units)

If the beam cross-section is considered to be circular, the moment of inertia is given as follows [5]:

$$I = \frac{\pi d^4}{64} \tag{2-5}$$

$I$ = *Moment of Inertia*
$d$ = *Diameter*

Using a and b dimensions of 4 in and 3 in respectively, a diameter of 0.25 in, a force of 10 lbf, and a modulus of elasticity of $10 \times 10^6$ psi, the capabilities of a symbolic equation solver to determine results with units is shown in Figure 2-3.

**Advantages of Symbolic Equation Solvers**

- Changes can be made easily and results evaluated quickly
- Since the equations are symbolic, they can be written to match published literature (including the use of Greek letters and other symbols)
- Allows concepts to be developed and derived without the errors associated with hand calculations

- May allow a closed-form solution to be developed for a specific situation

- Units can be included in the constants to help eliminate errors in conversions

**Disadvantages of Symbolic Equation Solvers**

- Extremely large equations may be difficult to solve due to hardware and/or software limitations

- Importing and exporting data with other applications may be limited

- Familiarity with software and/or computer equipment required

---

Beam bending energy equation:

$$U = \frac{1}{2 \cdot E \cdot I} \cdot \left[ \int_0^a (F \cdot x)^2 \, dx + \int_a^{a+b} (F \cdot a)^2 \, dx \right]$$

Evaluating symbolically:

$$U = \frac{1}{(2 \cdot (E \cdot I))} \cdot \left[ \frac{-2}{3} F^2 \cdot a^3 + F^2 \cdot a^2 \cdot (a+b) \right]$$

Taking partial derivative with respect to F:

$$\delta_F = \frac{1}{(2 \cdot (E \cdot I))} \cdot \left[ \frac{-4}{3} F \cdot a^3 + 2 \cdot F \cdot a^2 \cdot (a+b) \right]$$

Simplifying:

$$\delta_F = \frac{1}{3} F \cdot a^2 \cdot \frac{(a + 3 \cdot b)}{(E \cdot I)}$$

**Figure 2-2 - Output of Symbolic Equation Solver**

Dimensions, loads, and material properties:

$$a := 4 \cdot in$$

$$b := 3 \cdot in$$

$$d := 0.25 \cdot in \quad diameter$$

$$F := 10 \cdot lbf \quad applied\ force$$

$$E := 10 \cdot 10^6 \cdot psi \quad modulus\ of\ elasticity$$

Moment of inertia:

$$I := \frac{\pi \cdot d^4}{64}$$

$$I = 1.917 \cdot 10^{-4} \cdot in^4$$

Deflection:

$$\delta_F := \frac{1}{3} \cdot F \cdot a^2 \cdot \frac{(a + 3 \cdot b)}{(E \cdot I)}$$

$$\delta_F = 0.362 \cdot in$$

**Figure 2-3 - Use of Symbolic Equation Solver with Units**

### 2.2.3    Spreadsheets

**Description**

A spreadsheet is a software product that provides a tabular worksheet which performs rapid numerical calculations. A spreadsheet is typically broken up into cells with columns designated by letters and rows designated by numbers. Equations can be developed within spreadsheets that use the values contained in other cells to determine the results of the equation. Typically, spreadsheets also include the ability to graph results and iteratively solve for a particular goal. To some extent, spreadsheets may be considered to be the "backbone" of electronic packaging analysis.

## Typical Application of Spreadsheet

Equation 2-4, developed using the symbolic equation solver (see Figure 2-2) to represent the beam shown in Figure 2-1, may be implemented using a spreadsheet to determine the results for various dimensions and/or material properties. Equations 2-4 and 2-5 are incorporated into a spreadsheet using cell references in place of the equation variables (see Figure 2-4). The spreadsheet is solved numerically as shown in Figure 2-5. From Figure 2-4, it can be seen that the spreadsheet representation of equation 2-4 is somewhat difficult to follow, and may be a potential source of error. To help eliminate these errors, initial results from a new spreadsheet should be checked carefully using an alternate calculation method (hand calculations, symbolic equation solver, etc.).

|    | A     | B   | C      | D       | E      | F                | G                                      |
|----|-------|-----|--------|---------|--------|------------------|----------------------------------------|
| 1  | Force | Dimensions (in) |  |  | Modulus | Moment of Inertia | Deflection                            |
| 2  | F (lb) | a  | b      | Dia. (d) | E (psi) | I (in^4)        | (in)                                   |
| 3  | 10    | 4   | 1      | 0.25    | 1e+07  | =PI()*D3^4/64    | =1/3*A3*B3^2*(B3+3*C3)/(E3*F3)         |
| 4  | 10    | 4   | 2      | 0.25    | 1e+07  | =PI()*D4^4/64    | =1/3*A4*B4^2*(B4+3*C4)/(E4*F4)         |
| 5  | 10    | 4   | 3      | 0.25    | 1e+07  | =PI()*D5^4/64    | =1/3*A5*B5^2*(B5+3*C5)/(E5*F5)         |
| 6  | 10    | 4   | 4      | 0.25    | 1e+07  | =PI()*D6^4/64    | =1/3*A6*B6^2*(B6+3*C6)/(E6*F6)         |
| 7  | 10    | 2   | 1      | 0.1875  | 1e+07  | =PI()*D7^4/64    | =1/3*A7*B7^2*(B7+3*C7)/(E7*F7)         |
| 8  | 10    | 2   | 2      | 0.1875  | 1e+07  | =PI()*D8^4/64    | =1/3*A8*B8^2*(B8+3*C8)/(E8*F8)         |
| 9  | 10    | 2   | 3      | 0.1875  | 1e+07  | =PI()*D9^4/64    | =1/3*A9*B9^2*(B9+3*C9)/(E9*F9)         |
| 10 | 10    | 2   | 4      | 0.1875  | 1e+07  | =PI()*D10^4/64   | =1/3*A10*B10^2*(B10+3*C10)/(E10*F10)   |

## Figure 2-4 - Spreadsheet Equations

|    | A     | B   | C      | D       | E      | F                | G          |
|----|-------|-----|--------|---------|--------|------------------|------------|
| 1  | Force | Dimensions (in) |  |  | Modulus | Moment of Inertia | Deflection |
| 2  | F (lb) | a  | b      | Dia. (d) | E (psi) | I (in^4)        | (in)       |
| 3  | 10    | 4   | 1      | 0.25    | 1e+07  | 1.917e-04        | 0.195      |
| 4  | 10    | 4   | 2      | 0.25    | 1e+07  | 1.917e-04        | 0.278      |
| 5  | 10    | 4   | 3      | 0.25    | 1e+07  | 1.917e-04        | 0.362      |
| 6  | 10    | 4   | 4      | 0.25    | 1e+07  | 1.917e-04        | 0.445      |
| 7  | 10    | 2   | 1      | 0.1875  | 1e+07  | 6.067e-05        | 0.110      |
| 8  | 10    | 2   | 2      | 0.1875  | 1e+07  | 6.067e-05        | 0.176      |
| 9  | 10    | 2   | 3      | 0.1875  | 1e+07  | 6.067e-05        | 0.242      |
| 10 | 10    | 2   | 4      | 0.1875  | 1e+07  | 6.067e-05        | 0.308      |

## Figure 2-5 - Spreadsheet Results

## Advantages of Spreadsheets

- Changes can be made easily and results evaluated quickly

- Many calculations can be performed simultaneously

- Tabular form is compatible with reports and presentations

- Ability to import and export data to other software applications

**Disadvantages of Spreadsheets**

- Use of cell references in complex equations can sometimes be cryptic

- Unit conversions must be tracked manually

- Familiarity with software and/or computer equipment required

## 2.2.4    Finite-Element Analysis

**Description**

Finite-element analysis is a method that solves for complex geometries by using simultaneous equations on a large number of simple geometries. Instead of analyzing a continuum as is often found in solid mechanics and thermal problems, a finite number of simple geometries or "elements" are used (see Figure 2-6). These elements are connected to adjacent elements at fixed points called nodes. Finite-element software varies between software vendors but most include the capability to analyze the following areas:

- Static mechanical

- Dynamic mechanical (shock) response

- Resonant frequencies and modeshapes

- Harmonic response

- Steady-state thermal

- Transient thermal

Many finite-element software packages include pre-processors to aid in the development of the nodes and elements in the mesh. Some pre-processors include solid model representations and the capability to import IGES (Initial Graphics Exchange Specification) files and other standardized geometric formats. Finite-element software also typically includes post-processors to present results in an understandable form.

**Specialized Finite-Element Analysis Software**

Specialized finite-element packages are available that include computational fluid dynamics (CFD) and non-linear mechanics (yielding, creep, etc.). Although specific applications of these packages are beyond the scope of this book, potential areas of use include convection heat transfer and solder-life analysis.

**Typical Application of Finite-Element Analysis**

The deflection of the angled beam shown in Figure 2-1 may be determined using finite-element analysis. Using a and b dimensions of 4 in and 3 in respectively and

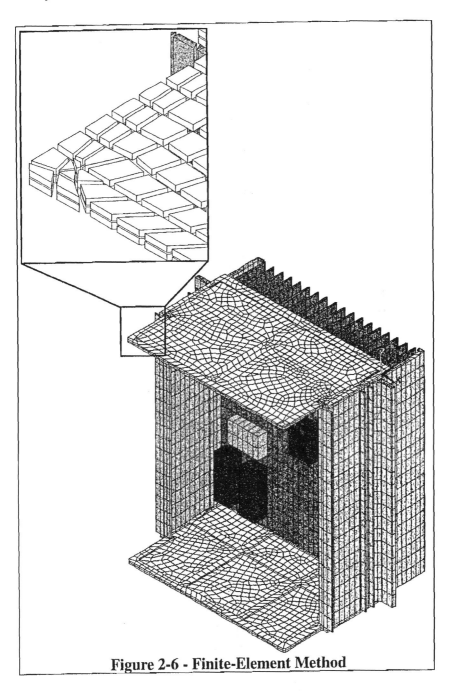

**Figure 2-6 - Finite-Element Method**

a diameter of 0.25 in (see Figure 2-3), a finite-element mesh of 2,108 nodes and 1,650 elements (see Figure 2-7) is developed. Applying a 10 lbf load (F) as shown in Figure 2-1, the deflection in the direction of the load is determined to be 0.369 in (see Figure 2-8). This agrees with the result obtained with equation 2-4 (see Figure 2-5) within 2%.

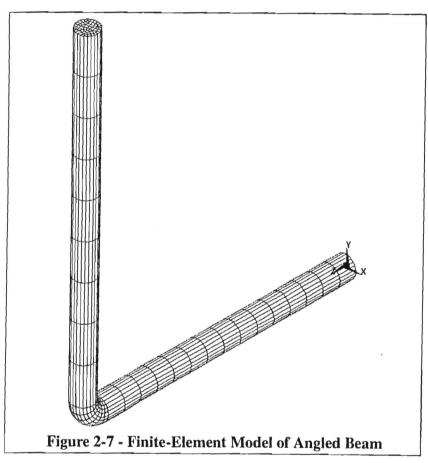

**Figure 2-7 - Finite-Element Model of Angled Beam**

### Advantages of Finite-Element Analysis

- Applicable to a very large range of geometries

- Finite-element solutions are possible in situations where closed-form analytical solutions are not possible or practical

- Ability to import data from computer-aided-design (CAD) software

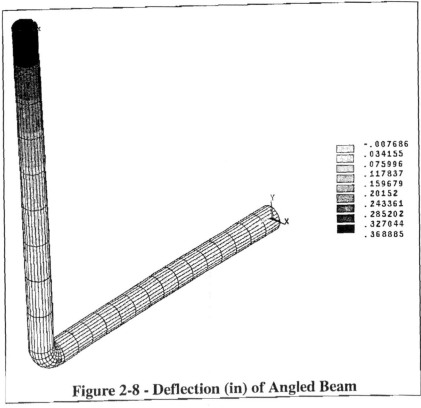

**Figure 2-8 - Deflection (in) of Angled Beam**

**Disadvantages of Finite-Element Analysis**

- Development of finite-element models may be labor-intensive
- Solution of finite-element models may require considerable computer resources
- Unit conversions must be tracked manually
- Familiarity with the finite-element method, software, and computer equipment required

## 2.2.5   Custom Programs

**Description**

A custom program is a computer program developed by the analyst (using C [6], FORTRAN [7], or other codes) to solve a particular problem or class of problems. Since the custom program is developed for a specific application, the limitations noted in some of the other analytical tools may not exist. The only limitations in a custom program are the capabilities of the hardware or software, or the ingenuity of the programmer. It important that any custom program be properly documented and/or commented so that others can understand the approach used. It is also important that information on revisions to the program be included. Since a user-developed program is subject to "bugs" and errors, verification of the results for a variety of input conditions must be conducted (see 8.3).

**Typical Custom Program**

A custom program used to solve the angled beam shown in Figure 2-1 is shown in Figure 2-9. This C programming language [6] program uses equations 2-4 and 2-5 to determine the deflection and moment of inertia. Results obtained by running the program closely match those obtained by other methods as shown in Figure 2-10.

```
/* Deflection of angled beam
   S.McKeown
   9/12/97
*/
#define VERSION "1.0"

#include <stdio.h>
#include <math.h>
#define PI 3.14159

main()
{
/* Declarations */
       float a,b,d,frc,delta,emod,iner;
/* Initial Information */
       printf("Deflection of Angled Beam (version %s)\n\n", VER-
SION);
/* Input Dimensions */
       printf("a Dimension (in)?\n");
       scanf("%f", &a);
       printf("b Dimension (in)?\n");
       scanf("%f", &b);
       printf("Diameter (d, in)?\n");
       scanf("%f", &d);
/* Input Applied Force */
       printf("Force (F, lbf)?\n");
       scanf("%f", &frc);
/* Input Material Properties */
       printf("Modulus of Elasticity (E, psi)?\n");
       scanf("%f", &emod);
/* Calculate and Print Moment of Inertia (equation 2-5) */
       iner=PI*pow(d,4)/64.0;
       printf("\nMoment of Inertia = %.3e in^4\n", iner);
/* Calculate and Print Deflection */
       delta=1/(3*emod*iner)*frc*pow(a,2)*(a+3*b);
       printf("\nDeflection = %.3f in\n\n", delta);
}
```

**Figure 2-9 - Typical Custom Program (C language)**

```
Deflection of Angled Beam (version 1.0)

a Dimension (in)?
4
b Dimension (in)?
3
Diameter (d, in)?
0.25
Force (F, lbf)?
10
Modulus of Elasticity (E, psi)?
10e6

Moment of Inertia = 1.917e-04 in^4

Deflection = 0.362 in
```

**Figure 2-10 - Custom Program Results**

**Advantages of Custom Programs**

- Portability between different computer platforms if standardized code is used

- Use of multiple-letter variables allows Greek and other special characters to be spelled out, allowing equations to be similar to published literature

- Applicable to almost any problem

- May be more efficient for some applications

**Disadvantages of Custom Programs**

- Development of custom program may be time-consuming (especially in the development of graphic and user-friendly applications)

- Subject to "bugs" and errors in programming

- Unit conversions, equation solutions, etc. are the responsibility of the programmer

- Familiarity with programming and computer equipment required

# 2.3   SUMMARY OF ANALYTICAL TOOLS

As mentioned before, selection of the appropriate analytical tool depends upon the specific situation and resources available. Relative advantages and disadvantages of various analytical tools are summarized in Table 2-1.

## Table 2-1 - Comparison of Analytical Tools

| Tool | Advantages | Disadvantages |
|------|------------|---------------|
| Hand Calculations | • No special tools required<br>• Familiarity with software or computer not required<br>• Concepts can be developed rapidly | • Difficult to make changes<br>• Difficult to check<br>• Subject to human errors<br>• Manual unit conversions<br>• Large volumes of data can be tedious and/or time-consuming |
| Symbolic Equation Solver | • Changes can be made easily and evaluated quickly<br>• Equations can match published literature<br>• Eliminates hand calculation errors<br>• May allow closed-form solutions<br>• Units can be included | • Extremely large equations may be difficult to solve<br>• Importing and exporting of data may be limited<br>• Familiarity with software and/or computer required |
| Spreadsheets | • Changes can be made easily and evaluated quickly<br>• Many calculations can be performed simultaneously<br>• Tabular form compatible with reports, etc.<br>• Can import and export data | • Cell references in complex equations can be cryptic<br>• Manual unit conversions<br>• Familiarity with software and/or computer required |
| Finite-Element Analysis | • Applicable to many geometries<br>• Can be used where closed-form solutions are not possible/practical<br>• Can import CAD data | • Development of models may be labor-intensive<br>• May require considerable computer resources<br>• Manual unit conversions<br>• Familiarity with software and/or computer required |
| Custom Programs | • Portability if standardized code is used<br>• Multiple-letter variables allow Greek and other special characters to be spelled out<br>• Applicable to almost any problem<br>• May be more efficient for some applications | • Development of program may be time-consuming<br>• Subject to "bugs" and errors in programming<br>• Unit conversions, equation solutions, etc. responsibility of the programmer<br>• Familiarity with programming and computer equipment required |

## 2.4    "SEAMLESS" AUTOMATION

### 2.4.1    Background

The almost universal use of computers in engineering has led to the development of various software packages and data formats. The analytical tools described in this chapter are only a small subset of the software that is available for various applications, including some packages specifically designed to perform electronic packaging analysis (such as CALCE [8]). With the prevalence of different software applications and platforms, the ability to transfer data between applications with a minimum of human interaction ("seamless" automation) is extremely important. Some software packages inherently include the ability to import and export data to other applications. In cases where import/export functions do not meet the specific requirements, custom programs and macros may be used.

### 2.4.2    Custom Programs and Macros

The use of a custom program to perform analytical calculations was described earlier (see 2.2.5) but custom programs may also be very helpful in transferring data from one application to another. A custom program for data transfer would read the data file format from one application and write it in a manner that is readable by another application. In some cases, data transfer and analysis can be combined in a single program that reads data from different sources and performs the required calculations. One possible application is the reading of printed wiring board (PWB) design data, finite-element thermal cycling and vibration analysis results, component data, and environment information to calculate solder-joint life [9] (see Figure 2-11). In some cases a software package may include the ability to record keystrokes in a macro to import and/or export data from another application. In other cases, custom program-like statements may be included in an application program input stream to generate data files that can be read by other applications.

## 2.5    ANALYTICAL TOOL CHECKLIST

### 2.5.1    Applicable to Hand Calculations

☐    Have the results been checked?

☐    Has the system of units been checked for compatibility?

### 2.5.2    Applicable to Symbolic Math Solvers

☐    Have the equations used in developing the solution been checked against source material?

```
Surface mount technology life (smtlife) Version 2.10

MIXED TECHNOLOGY DATA:

17 types read from mixed technology file

PWB LOCATION DATA:

Ref.         Component      X          Y        Power    X        Y
Des.           Type      Location  Location    (mW)    Dim.     Dim.   Pads Dir.
C150         cdr04_mxt     1.598     7.730       0    0.1800   0.1250    2   -
.

R965         rm1005_b_mxt  2.064     1.990       1    0.1000   0.0500    2   -

257 components oriented in X-direction
0 components in Y-direction
257 components read from location file

PWB:  Kevlar/epoxy
Heat Sink:  Aluminum
Bond:  Silicone

Bond thickness = 0.015 in
5513 CTE data points read
PWB size 4.675 x 9.518, reference at (2.173,9.138)
X-range = -0.165 to 4.510
Y-range = -0.380 to 9.138
X CTE = 9.17 to 11.29
Y CTE = 9.63 to 13.36

VIBRATION DATA:

Heat Sink:  Aluminum

Bond:  Silicone

PWB thickness = 0.057 in
Bond thickness = 0.015 in
Frequency = 228.81 Hz
Reference PSD = 0.04 g^2/Hz
133 vibration data points read
PWB size 4.680 x 10.063, reference at (2.640,9.463)
X-range = 0.300 to 4.980
Y-range = -0.600 to 9.463
Vibration reference adjusted by -0.4675 (X) and -0.3250 (Y)

Mission Environment

THERMAL MISSIONS:

Mission 1
         Code: 94K
         Maximum Temperature = 40.4 deg C
         Minimum Temperature = -9.9 deg C
         Power Fraction = 100%
         Dwell Time = 23.50 min
         Mission Duration = 1.567 hr
         Number Missions = 9574
.

Total Duration of All Thermal Missions = 50011 hr

VIBRATION MISSIONS:

Mission 1
         Code: ESS
         PSD = 0.0400 g^2/Hz
         Duration = 0.250 hr

Total Duration of All Vibration Missions = 0.250 hr

LIFE ANALYSIS:

Ref.         Component      Vibration  Thermal  Cumulative
Des.           Type         Factor     Factor   Fail.Prob.
C150         cdr04_mxt       0.0133     0.3078    0.02%
.

R965         rm1005_b_mxt    0.0002     0.5326    0.05%
```

**Figure 2-11 - Output of Solder-Life Custom Program**

### 2.5.3     Applicable to Spreadsheets

☐   Have the initial results of a new spreadsheet been checked?

☐   Has the system of units been checked for compatibility?

### 2.5.4     Applicable to Finite-Element Analysis

☐   Has the input data and mesh size been verified?

☐   Has the system of units been checked for compatibility?

### 2.5.5     Applicable to Custom Programs

☐   Have the analytical method and relevant equations been checked for accuracy?

☐   Has the system of units been checked for compatibility?

☐   Has the custom program been checked for "bugs" and errors?

☐   Have the appropriate documentation and/or comments been included?

☐   Has the revision history been included?

☐   Has the version number been included in the program output?

## 2.6     REFERENCES

[5]   Volterra & Gaines; Advanced Strength of Materials; Prentice-Hall, 1971

[6]   Kernighan & Ritchie; The C Programming Language; Prentice-Hall, 1988

[7]   Etter, D.M.; Structured FORTRAN 77 for Engineers and Scientists, Third Edition; Benjamin/Cummings Publishing Company, Inc., 1990

[8]   Computer-Aided Life-Cycle Engineering (CALCE) Electronic Packaging Research Center, University of Maryland, College Park, MD

[9]   McKeown, S.; "Automated Solder Life Analysis of Electronic Modules", (presented at the 126th TMS Annual Meeting and Exhibition, Orlando, FL, February 10, 1997); Design and Reliability of Solders and Solder Interconnections; The Minerals, Metals & Materials Society, Warrendale, PA, 1997; pp 279-286

# Chapter 3
# Thermal Performance Analysis

## 3.1　BACKGROUND

The goal of thermal performance analysis is typically to determine junction temperatures of electronic components for reliability assessment and to ensure that manufacturer's ratings are met. An additional, but related, goal is to evaluate top-level thermal management for use in determining environmental and cooling requirements. Electronic packaging configurations are typically increasing in performance and speed while decreasing in size, resulting in greatly increased power density. This increased power density results in increased internal temperatures which can reduce reliability and life if heat is not removed from the hardware. Heat can be removed by conduction, convection, radiation or a combination of the three methods. Any thermal analysis of electronic packaging must determine the thermal path and the applicable heat transfer mechanisms. Thermal analysis of electronic packaging configurations can be broken down into two categories, steady-state and transient thermal analysis.

### 3.1.1　Steady-State Thermal Analysis

A steady-state thermal condition occurs when the equipment under consideration is in thermal equilibrium where the internal heat generation equals the heat dissipated to the environment and/or cooling medium. Since the equipment is in equilibrium, no further temperature changes will occur. This type of condition would arise if a piece of equipment were required to operate in a constant environment for several hours.

### 3.1.2　Transient Thermal Analysis

A transient thermal condition occurs when the equipment under consideration is not in thermal equilibrium and there is a difference between the internal heat generation and the heat dissipated to the environment and/or cooling medium. Equipment temperatures will continue to change until the equipment reaches thermal equilibrium. This type of condition would arise if a piece of equipment were required to operate for a certain period of time after a loss of coolant. Transient analysis is also required if the outside environment was expected to undergo rapid temperature changes.

## 3.2    SELECTION OF ANALYSIS APPROACH

The selection of the approach for thermal performance analysis depends upon the complexity of the design, accuracy required, and the type of result required. Classical methods usually require simplifications of geometry and other approximations but are useful for determining if potential problems exist, conducting trade-off studies, verifying results from other analyses, or developing an understanding of the thermal design. The classical analysis method may be automated by use of a spreadsheet, or symbolic equation solver. Finite-element methods can consider detailed geometrical or material variations but usually require more setup labor and/or computer resources. An underlying understanding of thermal design and relationships is important to properly verify the finite-element results. Chapter 2 provides additional information on the selection of the analytical approach.

## 3.3    THERMAL CONFIGURATION

### 3.3.1    Thermal Path

It is important to consider the effect of the thermal path at all levels of packaging and for various packaging configurations, as shown in Figure 3-1. If, for example, poor thermal contact exists between the component package and the component heat sink, the heat transfer may be dominated by the thermal path through the leads. In many cases two parallel thermal paths may be significant. To avoid inaccuracies it is recommended that, unless the thermal path is completely known, that all applicable heat flow paths be considered in any thermal analysis. Thermal paths that are insignificant will be readily identified as the analysis progresses. Once the appropriate heat flow paths have been identified, a thermal schematic (see Figure 3-2) can be developed using thermal resistances, thermal masses (see 3.3.2), and thermal boundary conditions (see 3.3.3). Although it is possible in some cases (such as finite-element analysis) to obtain a solution without a thermal schematic, the schematic contributes to the understanding of the thermal situation and may aid in validating/verifying results.

### 3.3.2    Thermal Masses

Thermal masses need only be considered in transient analyses, but it is important to properly account for all of the mass in the system. Proper accounting of mass

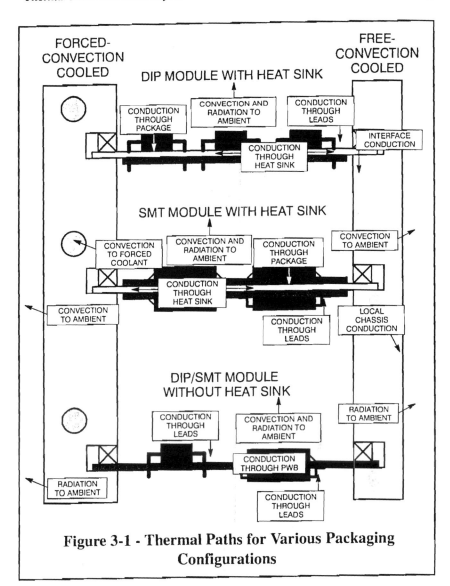

**Figure 3-1 - Thermal Paths for Various Packaging Configurations**

also requires determining which masses should be lumped or combined together and which should be separated.

### 3.3.3    Thermal Boundary Conditions

It is important to accurately determine the thermal boundary conditions applicable to the equipment under analysis. Specific thermal boundary conditions to be considered include the following:

- Power dissipation of components
- Ambient temperature
- Coolant temperature (if applicable)
- Temperature of surrounding equipment for radiation calculations

These thermal boundary conditions are usually defined by environmental requirements, electrical design, surrounding equipment design, or a combination of these or other methods. Once thermal boundary conditions have been defined, it is important to properly document the source of the thermal boundary conditions for future reference.

### 3.3.4    Packaging Levels

Electronic packaging can be divided into three levels. The first level includes the components that are mounted on the printed wiring boards (PWBs); the second level includes the modules or PWBs themselves; and the third level includes the chassis or enclosure into which the modules/PWBs are mounted. For the purposes of classical thermal analysis, the first packaging level includes thermal resistances between the component and average module temperature. The second level includes thermal resistances between the average module temperature and the average chassis temperature. The third level includes thermal resistances between the average chassis and the ambient and/or coolant. When finite-element methods are used, the packaging levels considered for thermal analysis more closely match the physical packaging levels.

## 3.4   COMPONENT (FIRST) LEVEL

### Type of Package

Heat transfer at the component level is significantly influenced by the type of package involved. This heat is usually conducted from the heat source (component die, resistive element, etc.) out through the body of the part and/or through the leads/interconnection as illustrated in Figure 3-3. This thermal path is controlled by the following factors:

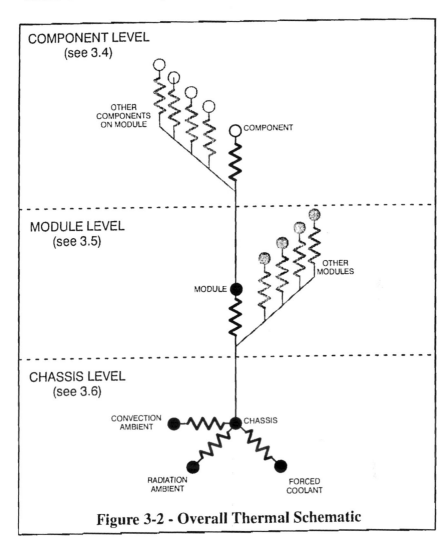

**Figure 3-2 - Overall Thermal Schematic**

- Package thermal resistance

- Use of heat spreaders

- Interconnection thermal resistance

- Degree-of-contact between component and heat sink and/or PWB

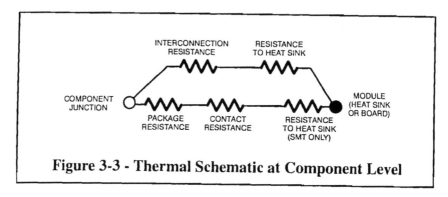

**Figure 3-3 - Thermal Schematic at Component Level**

## 3.4.1    Classical Methods

**Package Thermal Resistance/Heat Spreading**

Basic Relationships

Thermal resistance through the package depends upon the spreading of the heat
(see Figure 3-4) through the die and package to the heat sink. Usually a spreading
angle of 45° is assumed but this will vary depending upon the degree of heat sinking applied to the exterior of the package. If the heat sinking at the exterior of the
package is poor, a steeper spreading angle may be used whereas if heat sinking is
exceptionally good, a shallower spreading angle should be used. Finite-element
analysis may be used to determine the spreading angle. For a square junction with
constant temperature across the interface surfaces, the thermal resistance from the
junction to the heat sink is as follows:

$$\theta_{jc} = \int_{0}^{l_d} \frac{dx}{(k_d(a + 2x \tan\phi)^2)}$$

$$+ \int_{l_d}^{l_d+l_b} \frac{dx}{(k_b(a + 2x \tan\phi)^2)} + \int_{l_d+l_b}^{l_d+l_b+l_p} \frac{dx}{(k_p(a + 2x \tan\phi)^2)} \tag{3-1}$$

$$\theta_{jc} = \frac{l_d}{k_d a(a + 2l_d \tan\phi)} + \frac{l_b}{k_b(a + 2l_d \tan\phi)[a + 2(l_d + l_b) \tan\phi]}$$

$$+ \frac{l_p}{k_p[a + 2(l_d + l_b) \tan\phi][a + 2(l_d + l_b + l_p) \tan\phi]} \tag{3-2}$$

$\theta_{jc}$ = *Junction-to-Case Resistance*
$a$ = *Length/Width of Junction*
$x$ = *Integration Variable (distance from die)*
$\phi$ = *Spreading Angle (deg)*
$l_p, l_b, l_d$ = *Thickness of Package, Die bond and Die*
$k_p, k_b, k_d$ = *Thermal Conductivity of Package, Die bond and Die*

## Underlying Assumptions

The above equations are subject to the following assumptions:

- Uniform heating on surface of die

- Constant temperature on interface surfaces

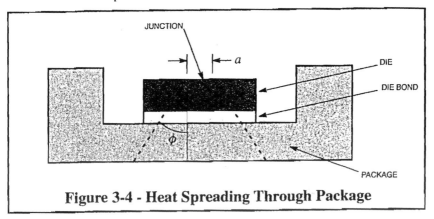

### Figure 3-4 - Heat Spreading Through Package

**Interconnection Thermal Resistance**

Thermal resistance of the interconnections between the package and PWB is determined by evaluating the thermal resistance of the electrical path between the package and board. This resistance is calculated differently if the component is leadless or leaded.

## Leadless Components

For leadless components the solder provides part of the heat path from the component to the PWB. Since the typical cross-section for a leadless chip-carrier solder joint is not uniform, the thermal resistance can be difficult to calculate. One way of obtaining a worst-case (maximum) value for the thermal resistance of the solder joints is to average the pad area on the printed wiring board (PWB) with the terminal area on the bottom of the chip-carrier. This average area is combined with the solder joint thickness to calculate the thermal resistance:

$$\theta_i = \frac{2l_s}{k_s\,(A_{pad} + A_{term})}$$

(3-3)

$\theta_i$ = Interconnection Resistance

$A_{pad}$ = Area of PWB Pad

$A_{term}$ = Area of Chip Carrier Terminal (on bottom of package)

$l_s$ = Thickness of Solder Joint

$k_s$ = Thermal Conductivity of Solder

This thermal resistance does not consider the contribution of the fillet which would tend to lower the thermal resistance. If a more accurate thermal resistance is required, the use of a finite-element model should be considered.

Leaded/DIP Components

For leaded components, a metal lead contributes to the heat path between the component package and the PWB. For surface-mount configurations this lead is usually straight so the thermal resistance is calculated somewhat easily. dual-inline-package (DIP) components usually include a short tapered area which must be considered in the thermal resistance calculation. Thermal resistance of a surface-mount lead is given as follows:

$$\theta_i = \frac{l_l}{k_l wt}$$

(3-4)

$\theta_i$ = Interconnection Resistance

$w$ = Width of Lead

$t$ = Thickness of Lead

$l_l$ = Length of Lead

$k_l$ = Thermal Conductivity of Lead

Thermal resistance of a DIP lead (with consideration given to tapered region) is similarly calculated:

$$\theta_i = \frac{l_1}{k_l\,w_1\,t} + \frac{l_l}{k_l\,(w_1 - w_2)\,t}\ln\!\left(\frac{w_1}{w_2}\right) + \frac{l_2}{k_l\,w_2\,t}$$

(3-5)

$\theta_i$ = *Interconnection Resistance*
$w_1$ , $w_2$ = *Widths of Lead*
$t$ = *Thickness of Lead*
$l_1$ , $l_2$ = *Lengths of Lead*
$l_t$ = *Length of Lead Transition Region*
$k_l$ = *Thermal Conductivity of Lead*

## Underlying Assumptions

- Uniform temperature of package in the area of lead/terminal
- Uniform temperature of PWB at terminal pad
- Contribution of fillet in leadless chip-carrier thermal resistance is neglected

## Thermal Resistance between Component and Module

### Thermal Contact between Component Body and Module

Thermal contact resistance between the component body and module will be different if a bonding material is applied in the interface from that expected for dry contact. If a bonding material is used the thermal resistance is that of a rectangular pad as follows:

$$\theta_c = \frac{t_b}{k_b \, A_c} \qquad (3\text{-}6)$$

$\theta_c$ = *Thermal Contact Resitance*
$t_b$ = *Thickness of Bond*
$A_c$ = *Area of Component (contact area)*
$k_b$ = *Thermal Conductivity of Bond*

For dry contact, the thermal contact between the component body and module is determined by dividing a thermal contact resistivity parameter by the area of contact to determine the thermal resistance as follows:

$$\theta_c = \frac{R_c}{A_c} \qquad (3\text{-}7)$$

$\theta_c$ = *Thermal Contact Resitance*
$R_c$ = *Thermal Contact Resistivity*
$A_c$ = *Area of Component (contact area)*

The thermal contact resistivity is determined from testing or published information. This contact resistivity would have units of temperature*area/power.

### Thermal Resistance to Heat Sink

In addition to the thermal contact resistance between the component and module, an additional temperature rise (see Figure 3-5) may be encountered due to the conduction of heat through the PWB and bond layer (in surface-mount configurations) or between the lead and heat sink (in through-hole configurations). The thermal resistance for a surface-mount configuration is simply the thermal resistance of a rectangular pad with the same dimensions as the component, with appropriate consideration given to the thickness and thermal conductivity for each layer as follows:

$$\theta_l = \frac{t_b}{k_b \, A_c} + \frac{t_{pwb}}{k_d \, A_d + k_{cu} \, A_{cu}} \tag{3-8}$$

$\theta_l = $ *Local Thermal Resistance between Component and Heat Sink*

$t_b, t_{pwb} = $ *Thickness of Bond and PWB*

$A_c, A_d, A_{cu} = $ *Area of Component, PWB Dielectric, and PTH Copper*

$k_b, k_d, k_{cu} = $ *Thermal Conductivity of Bond, PWB Dielectric, and Copper*

It is important to properly account for any plated-through holes (PTHs) that enhance thermal conductivity by determining the portion of the area allocated to such holes.

The thermal resistance from the lead to heat sink in a through-hole configuration is typically a complex thermal configuration and may require finite-element analysis. In many cases the direct contact between the component body and heat sink in a through-hole configuration dominates the heat transfer, so the heat transfer from the component through the lead and PWB to the heat sink may be neglected.

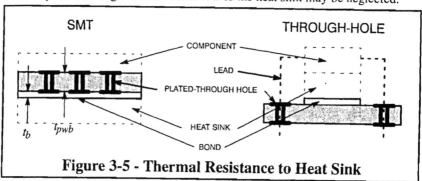

**Figure 3-5 - Thermal Resistance to Heat Sink**

### Convection and Radiation Loss from Components

Since electronic assemblies are typically closed, there are no significant temperature differences between the components and the ambient and/or the components and surrounding hardware. Since there is not a large temperature difference, the convection and radiation losses directly from the components are usually neglected, which gives a worst-case condition. In some applications, particularly those using plastic parts, cooling air is passed through the electronic enclosure by free or forced convection. In these situations, the convection and radiation loss may be significant and will need to be considered using the methods described in section 3.6.2. For consumer electronics applications, the convection loss from the components is a major effect which should be considered.

**Component Thermal Mass**

Background

Since components are usually small compared to the overall electronic assembly, the thermal mass of the components is typically small. The thermal mass of components is usually neglected in transient thermal analyses and the worst-case assumption of steady-state operation is used.

Underlying Assumptions

- Thermal mass of components is neglected

## 3.4.2    Finite-Element Analysis

Finite-element analysis may be used to calculate the thermal performance of a component. Usually the die, die bond, package, and leads are all included in a single finite-element model (see Figure 3-6). Thermal finite-element models usually include only one degree-of-freedom (temperature) at each node so the storage requirements and calculation times are typically less than that required for structural models. This allows additional model complexity to be included (if necessary) to provide the required accuracy.

## 3.4.3    Use of Published/Vendor Data

Vendors frequently publish thermal resistance values for their components. Usually these take the form of junction-to-case resistance for ceramic components, and junction-to-ambient values for plastic devices. It is important to understand the specific conditions that the vendor used to determine the thermal resistance [11]. This is especially true for plastic components which sometimes quote the junction-to-ambient values for the component mounted on a PWB. If the part is mounted on a board, part of the quoted thermal resistance is included at the second level of packaging. It is important to properly understand the situation to avoid "double counting" the effect of the PWB. In critical/marginal applications, published thermal resistance values should be verified by testing and/or analysis.

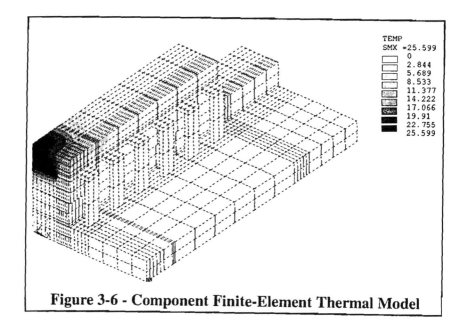

**Figure 3-6 - Component Finite-Element Thermal Model**

## 3.5    PWB/MODULE (SECOND) LEVEL

### 3.5.1    Identification of Thermal Path

Heat transfer at the module level is strongly influenced by the type of module package considered (see Figure 3-7). This heat transfer is usually by conduction from the heat component out through the PWB, or heat sink to the chassis rail as illustrated in Figure 3-8. This thermal path is controlled by the following factors:

- Thermal resistance of PWB

- Thermal resistance and configuration of heat sink

- Degree-of-contact between module and component rail

- Local temperature rise from module location to bulk chassis

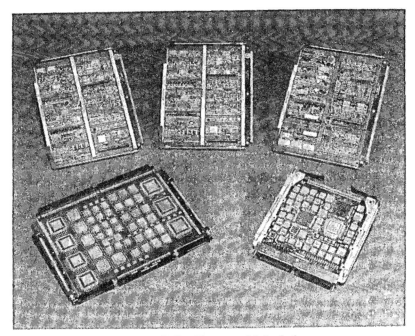

**Figure 3-7 - Types of Modules**
(Courtesy of Lockheed Martin Control Systems.)

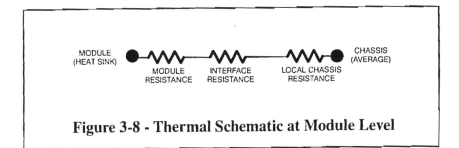

**Figure 3-8 - Thermal Schematic at Module Level**

## 3.5.2    Classical Methods

### Module Thermal Resistance

#### Through-Hole Technology

Thermal resistance of through-hole technology modules is principally governed by the resistance of the ladder heat sink. As heat is conducted out of the bottom of the DIP package, the heat travels into the heat sink and along the edges to the chassis rail. Heat sink rails can be considered thermal resistors between nodal points defined by component locations (see Figure 3-9). Resistance of these thermal resistors is calculated by a simple resistance of a prismatic bar as follows:

$$\theta_j = \frac{l_j}{k_{hs} \, w_j \, t} \tag{3-9}$$

$\theta_j$ = Resistance of Ladder Heat Sink Segment j
$w_j$ = Width of Ladder Heat Sink Segment j
$t$ = Thickness of Heat Sink
$l_j$ = Length of Ladder Heat Sink Segment j
$k_{hs}$ = Thermal Conductivity of Heat Sink

Once the individual resistances of each thermal resistor comprising the heat sink are determined, the overall thermal performance is determined by calculating the temperatures at each node of the resistor network. Proper determination of temperature depends upon the correct power being applied to each node. When networks are very complicated, finite-element analysis or a electrical network analysis tool may be required to calculate the temperatures.

#### Surface-Mount Technology

Thermal performance of a surface-mount technology heat sink is typically determined by assuming that the power dissipation over the surface of the module is uniform. If there are no large concentrated heat dissipators, the combination of the PWB and bond layer tends to spread the heat over the surface of the heat sink thus approximating a uniform heat distribution. Usually heat conducted into the surfaces is taken out by conduction through the length of the heat sink (see Figure 3-10). This results in a one-dimensional thermal relationship similar to a solid with internal heat generation. Both the average and maximum heat sink tempera-

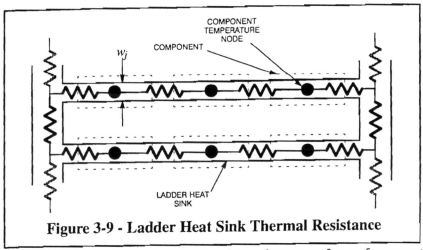

**Figure 3-9 - Ladder Heat Sink Thermal Resistance**

ture are useful in determining the thermal performance of a surface-mount
technology (SMT) module. These temperatures are calculated as follows:

$$\Delta T_x = P\frac{(l^2 - 4x^2)}{8k_{hs}lwt} \qquad (3\text{-}10)$$

$$\Delta T_c = \frac{Pl}{8k_{hs}wt} \qquad (3\text{-}11)$$

$$\Delta T_{avg} = \frac{Pl}{12k_{hs}wt} \qquad (3\text{-}12)$$

$P = Module\ Power$

$\Delta T_x = Heat\ Sink\ Temperature\ Rise\ above\ Edge\ at\ Distance\ x\ from\ Center$

$\Delta T_c = Heat\ Sink\ Center\ to\ Edge\ Temperature\ Rise$

$\Delta T_{avg} = Heat\ Sink\ Average\ to\ Edge\ Temperature\ Rise$

$w = Width\ of\ Heat\ Sink$

$t = Thickness\ of\ Heat\ Sink$

$l = Length/Span\ of\ Heat\ Sink$

$k_{hs} = Thermal\ Conductivity\ of\ Heat\ Sink$

Configurations without Heat Sinks

With some applications, particularly those using plastic parts, no metallic heat
sink is used. In these configurations, a considerable fraction of the heat is conducted through the component leads into the PWB. Since the components represent a localized power dissipation and the heat-spreading effect of the PWB over
the bond layer is no longer applicable, a simplified solution is not usually possible.

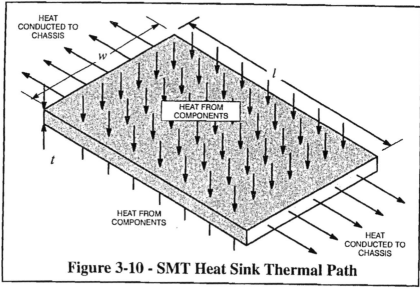

HEAT
CONDUCTED TO
CHASSIS

*w*

*l*

HEAT FROM
COMPONENTS

*t*

HEAT FROM
COMPONENTS

HEAT
CONDUCTED TO
CHASSIS

**Figure 3-10 - SMT Heat Sink Thermal Path**

Typically these configurations require finite-element analysis to account for non-uniformity in PWB conduction and localized power dissipation. Rough temperature rises can be determined by using the relationships for the SMT heat sink with the appropriate PWB properties used in place of the heat sink properties. Alternatively, a combination of the finite-element and rough analysis methods may be used. In this approach the finite-element method is used to develop an effective thermal conductivity of the board which is then used in the SMT heat sink relationships. In some cases, convection from the surface of the component may need to be considered.

<u>Underlying Assumptions</u>

- Uniform power distribution over surface of SMT heat sink

- Uniform temperature distribution across surface of PWB and bond line under SMT component

**Module-to-Chassis Interface**

Thermal contact between the module and chassis is determined by using equation 3-7 with the appropriate thermal contact resistivity parameter and the area of contact between the module and chassis. It is important to consider the interface pressure as influenced by the module clamping method in obtaining the thermal contact resistivity parameter.

### Local Chassis Thermal Resistance

Since an electronic chassis typically includes a relatively complex geometry, finite-element analysis is typically more appropriate for calculating local chassis thermal resistance (see 3.6.3). Approximate consideration of thermal resistance of the module rails and simplifications of chassis geometry (see Figure 3-11) may be appropriate by adapting the thermal resistance of a prismatic bar as follows:

$$\theta = \frac{l}{k \; w \; 2t} \tag{3-13}$$

$\theta = Thermal\ Resistance$

$w = Width$

$t = Thickness$

$l = Length$

$k = Thermal\ Conductivity$

The 2 in the denominator of equation 3-13 is necessary because heat is transferred through both edges of the module. Test data and/or finite-element analysis may also be used to calculate thermal resistances for use in classical models.

### Module Thermal Mass

Background

As described in 3.3.2, thermal masses need only be considered in cases where transient thermal analysis is required. In these cases, the thermal mass is determined by summing the product of the mass of each part (PWB, heat sink, component, etc.) by the appropriate specific heat.

$$Mc_{mod} = \Sigma \; m_i \; c_i \tag{3-14}$$

$Mc_{mod} = Total\ Thermal\ Mass$

$m_i = Mass\ of\ Part\ i$

$c_i = Specific\ Heat\ of\ Part\ i$

It important to sum the thermal masses in a method consistent with the thermal resistance calculations. Although it may be appropriate to use one thermal mass for a module which is assumed to be at some average temperature, with individual component temperatures taken as a difference from this average temperature, it is important to consider individual thermal masses/temperatures if a large concentrated thermal mass is present.

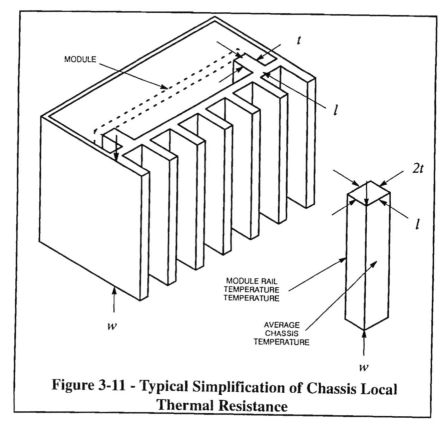

**Figure 3-11 - Typical Simplification of Chassis Local Thermal Resistance**

Underlying Assumptions

- Concentrated thermal masses are at a uniform average temperature

### 3.5.3    Finite-Element Analysis

In cases where geometry, power dissipation, or boundary conditions are non-uniform; a finite-element analysis is typically required. Finite-element analysis may also be necessary where the thermal masses cannot be adequately described by a lumped masses. As described in 3.4.2, thermal finite-element models typically include only one degree-of-freedom per node which allows greater modeling detail with given computer resources than with structural models. Finite-element methods may actually be easier for modeling ladder heat sinks used on DIP modules (see Figure 3-12) than the thermal resistor method described previously. Finite-element analysis is especially appropriate if an electronic method is available for describing the module geometry such as an IGES (Initial Graphics Exchange Specification) file. Thermal finite-element analysis at the module level may make use of substructures or super-elements to model details that are replicated such as those found in the internal construction of PWBs.

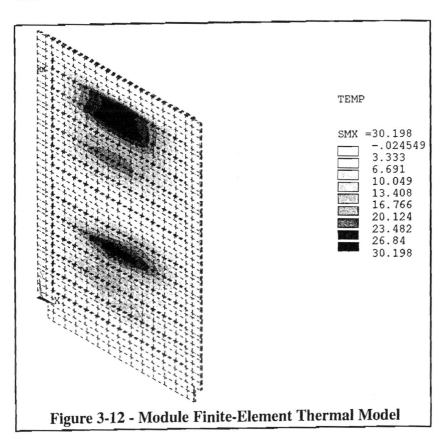

TEMP

SMX =30.198
-.024549
3.333
6.691
10.049
13.408
16.766
20.124
23.482
26.84
30.198

**Figure 3-12 - Module Finite-Element Thermal Model**

# 3.6    CHASSIS (THIRD) LEVEL

## 3.6.1    Identification of Thermal Path

Heat transfer at the chassis level is usually by free and/or forced convection and radiation from the chassis to the ambient/coolant as illustrated in Figure 3-13. This thermal path is controlled by the following factors:

- Coolant fluid and flow rate (where applicable)
- Chassis temperature
- Use of fins on outside of chassis
- Chassis size

Since free convection and radiation are usually non-linear in temperature, the heat losses from the chassis to the ambient are better expressed as heat loss rates instead of thermal resistances.

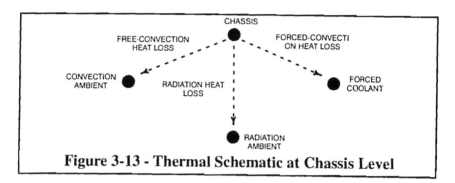

**Figure 3-13 - Thermal Schematic at Chassis Level**

## 3.6.2    Classical Methods

**Convection Boundary Conditions**

Forced Convection

Forced convection in electronic equipment typically uses air as the cooling medium but may use other gases or liquids as requirements dictate or as available. Temperature performance in forced-convection situations includes the contribution of both the bulk temperature rise of the fluid and the film temperature rise between the chassis and the fluid. If flow rates are high such that the fluid temperature rise is small compared to the film temperature rise, the average bulk fluid temperature may be used to determine the thermal performance as follows:

$$q_{forced} = (T_{chas} - T_{in})\frac{2\dot{m}c_{pf}h_fA_{fc}}{(2\dot{m}c_{pf} + h_fA_{fc})} \tag{3-15}$$

$$T_x = T_{in} + \frac{q_{forced}}{2\dot{m}c_{pf}} \tag{3-16}$$

$$T_{out} = T_{in} + \frac{q_{forced}}{\dot{m}c_{pf}} \tag{3-17}$$

$q_{forced}$ = *Forced Convection Heat Transfer Rate*
$T_{chas}$ = *Chassis Temperature*
$T_{in}$ = *Coolant Inlet Temperature*
$\dot{m}$ = *Coolant Flow Rate*
$c_{pf}$ = *Specific Heat of Cooling Fluid*
$h_f$ = *Convection Coefficient to Fluid*
$A_{fc}$ = *Area of Fluid to Chassis Interface*
$T_{\infty}$ = *Coolant Free Stream Temperature (for convection calculations)*
$T_{out}$ = *Coolant Outlet Temperature*

If the fluid temperature rise is large compared to the film temperature rise, as would be expected at low flow rates, the thermal calculations are more complex:

Differential Equation:

$$\frac{dT}{dx} = \frac{h_f \, p}{\dot{m} c_{pf}}(T_{chas} - T(x)) \tag{3-18}$$

Solution to Differential Equation:

$$T(x) = (T_{in} - T_{chas}) \, e^{-\left(\frac{h_f p x}{\dot{m} c_{pf}}\right)} + T_{chas} \tag{3-19}$$

Resulting Thermal Parameters:

$$T_{out} = (T_{in} - T_{chas}) \, e^{-\left(\frac{h_f A_{fc}}{\dot{m} c_{pf}}\right)} + T_{chas} \tag{3-20}$$

$$q_{forced} = \dot{m} c_p \left[ 1 - e^{-\left(\frac{h_f A_{fc}}{\dot{m} c_{pf}}\right)} \right](T_{chas} - T_{in}) \tag{3-21}$$

$h_f$ = *Convection Coefficient to Fluid*
$p$ = *Perimeter of Cooling Path Cross Section*
$\dot{m}$ = *Coolant Flow Rate*
$c_{pf}$ = *Specific Heat of Cooling Fluid*
$T_{chas}$ = *Chassis Temperature*
$T(x)$ = *Fluid Temperature at Distance x*
$T_{in}$ = *Coolant Inlet Temperature*
$A_{fc}$ = *Area of Fluid to Chassis Interface*
$T_{out}$ = *Coolant Outlet Temperature*
$q_{forced}$ = *Forced Convection Heat Transfer Rate*

## Table 3-1 - Equation 3-22 Constants for Fully Turbulent Forced Convection

| Condition | Pr | Re | C | m | n | Characteristic Dimension (a) |
|-----------|-----|-----|-----|-----|-----|------------------------------|
| Over Chassis | 0.5–1.0 | >500,000 | 0.0369 | 0.8 | 0.6 | Length parallel to flow |
| Tube [10] | 0.5–1.0 | >2,300 | 0.0210 | 0.8 | 0.6 | Hydraulic Diameter (4A/P) |
| Tube [10] | 1.0–20 | >2,300 | 0.0155 | 0.83 | 0.5 | |
| Tube [10] | >20 | >2,300 | 0.0118 | 0.9 | 0.3 | |

Values for the convection coefficient to the fluid are obtained from measurements or theoretical forced-convection relationships. Typically the convection coefficient is based upon the Nusselt number which is calculated as a function of the Reynolds and Prandtl numbers:

$$Nu_a = C\, Re_a{}^m Pr^n \tag{3-22}$$

$Nu_a = \dfrac{h\,a}{k} = Nusselt\ Number\ (dimensionless)$

$Re_a = \dfrac{\rho\,a\,v}{\mu} = Reynolds\ Number\ (dimensionless)$

$Pr = \dfrac{c_p\,\mu}{k} = Prandtl\ Number\ (dimensionless)$

$C,\ m,\ n = Constants\ (see\ text\ and\ Table\ 3\text{-}1)$

$h = Convection\ Coefficient$

$a = Characteristic\ Dimension\ (see\ text\ and\ Table\ 3\text{-}1)$

$k = Fluid\ Thermal\ Conductivity$

$\rho = Fluid\ Density$

$v = Coolant\ Velocity$

$\mu = Absolute\ Viscosity$

$c_p = Fluid\ Specific\ Heat$

Typical values for the constants used in equation 3-22 for turbulent forced convection in a cooling tube or over the surface of a heated electronic chassis are listed in Table 3-1. The characteristic dimension (a) for the above relationships is the length parallel to flow for flow over a chassis, and the hydraulic diameter for flow inside a tube or channel. The hydraulic diameter is defined as four times the flow area divided by the perimeter (4A/P) which is the diameter for circular tubes. Values for the Nusselt number for other configurations such as flow over cylinders, in finned plenums, or laminar forced convection may be found in heat-transfer texts or other reference material.

## Free Convection

Free convection occurs when the fluid is driven by density changes due to the heating process. For an electronic enclosure, typically the convection coefficient is higher from vertical surfaces and lower from horizontal surfaces (especially the bottom of the enclosure). Since the convection coefficient varies over the surface of the enclosure, it is important to consider the top, bottom, and sides separately. The convection coefficient also increases with increasing temperature which makes the model non-linear, so iteration on chassis temperature is sometimes required. Thermal resistance from chassis to ambient is represented as follows:

$$q_{free} = (h_s A_s + h_t A_t + h_b A_b + h_e A_e + h_f \eta_f A_f)\ (T_{chas} - T_\infty) \qquad (3\text{-}23)$$

$q_{free}$ = *Free Convection Heat Transfer Rate*

$h_s, h_t, h_b, h_e$ = *Convection Coefficient on Sides, Top, Bottom, Ends*

$A_s, A_t, A_b, A_e$ = *Area of Sides, Top, Bottom, Ends (excluding fins)*

$A_f$ = *Surface Area of Fins*

$h_f$ = *Convection Coeffcient for Finned Surfaces (see text)*

$\eta_f$ = *Fin Efficiency (see text)*

$T_{chas}$ = *Chassis Temperature*

$T_\infty$ = *Ambient Temperature*

Values for the free-convection coefficient may be determined by measurements or theoretical relationship. The Nusselt number for free convection is calculated as a function of the Grashof and Prandtl numbers:

$$h = \frac{Nu_a\ k}{a} \qquad (3\text{-}24)$$

$$Nu_a = C(Gr_a\ Pr)^m \qquad (3\text{-}25)$$

$$Gr_a = \frac{g\beta(T_{chas} - T_\infty)a^3\rho^2}{\mu^2} \qquad (3\text{-}26)$$

$Nu_a$ = Nusselt Number (dimensionless)

C, m = Constants (see text and Table 3-2)

$Gr_a$ = Grashof Number (dimensionless)

$Pr = \dfrac{c_p \mu}{k}$ = Prandtl Number (dimensionless)

h = Convection Coefficient

a = Characteristic Dimension (see text and Table 3-2)

k = Fluid Thermal Conductivity

g = Acceleration Due to Gravity

$\beta$ = Volume Coefficient of Expansion (constant pressure) $\left(\dfrac{1}{V}\left(\dfrac{\partial V}{\partial T}\right)_p\right)$

$T_{chas}$ = Chassis Temperature

$T_\infty$ = Ambient Temperature

$\rho$ = Fluid Density

$\mu$ = Absolute Viscosity

$c_p$ = Fluid Specific Heat

Note: Material properties evaluated at film temperature $\left(\dfrac{T_\infty + T_{chas}}{2}\right)$

Typical values for the constants used in equation 3-25 for laminar free convection from the surfaces of a heated electronic chassis are listed in Table 3-2. The characteristic dimension (a) for the above relationships is the height of the vertical surfaces or the mean of the sides for rectangular horizontal surfaces. Values for the Nusselt number for other configurations (such as cylinders) or for turbulent free convection may be found in heat-transfer texts or other reference material. If the sides are not vertical, appropriate considerations must be made in the calculation of the heat transfer coefficient. Since the Grashof number depends upon density, it is also important to consider the effect of altitude (ambient pressure) on the heat transfer coefficient. When fins are present, it is important to include the consideration of the fin proximity in the free-convection coefficient calculations for both the fins and the base area of the surfaces to which the fins are attached. It is also important to subtract the area of the chassis covered by the base of the fins for the finned chassis surfaces.

## Finned Structures

Fins may be used to reduce the convection thermal resistance by increasing the surface area, but the reduction in thermal resistance is not as much as the increase in surface area for the following reasons:

- Temperature drop along the fin reduces the heat lost from the fin
- Convection coefficient is reduced by the effect of the adjacent fin

The effect of temperature drop along the fin leads to the concept of fin efficiency which represents the ratio of the actual heat transferred to the heat transferred if the fin was at a constant temperature. For a fin with heat transfer from the ends (see Figure 3-14), the following relationship gives the fin efficiency:

$$\eta_f = \frac{\tanh(mL_c)}{mL_c} \tag{3-27}$$

$$L_c = L_f + \frac{t_f}{2} \tag{3-28}$$

$$m = \sqrt{\frac{h_f \, p_f}{k_f \, A_f}} \tag{3-29}$$

$$p_f = 2(H + t_f) \tag{3-30}$$

$$A_f = H \, t_f \tag{3-31}$$

$\eta_f$ = Fin Efficiency
$L_c$ = Effective Fin Length
$L_f$ = Actual Fin Length
$t_f$ = Fin Thickness
$m$ = Fin Characteristic Constant
$h_f$ = Convection Coefficient Surrounding Fin
$p_f$ = Fin Perimeter
$k_f$ = Fin Thermal Conductivity
$A_f$ = Fin Cross Sectional Area
$H$ = Fin Height

## Table 3-2 - Equation 3-25 Constants for Laminar Free Convection from a Heated Chassis [12]

| Surface | $Gr_aPr$ | C | m | Characteristic Dimension (a) |
|---------|----------|------|-------|------------------------------|
| Top | $10^5 - 2\times10^7$ | 0.54 | 0.25 | Mean of sides |
| Top | $2\times10^7 - 3\times10^{10}$ | 0.14 | 0.333 | Mean of sides |
| Bottom | $3\times10^5 - 3\times10^{10}$ | 0.27 | 0.25 | Mean of sides |
| Side | $10^4 - 10^9$ | 0.59 | 0.25 | Height |

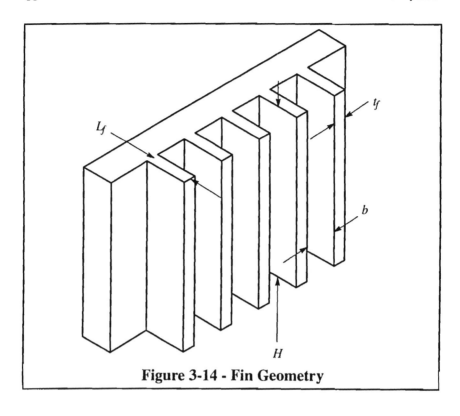

**Figure 3-14 - Fin Geometry**

The effect of fin spacing is a complex relationship but an excellent formulation for two-dimensional laminar flow between parallel plates may be found in the GE Heat Transfer Data Book [13]:

$$h_f = \frac{Nu_b \, k}{b} \tag{3-32}$$

$$Nu_b = \left[ \frac{576}{(Ra')^2} + \frac{2.87}{\sqrt{Ra'}} \right]^{-\frac{1}{2}} \tag{3-33}$$

$$Ra' = \frac{\rho^2 \, c_p \, g\beta \, b^4 (T_{chas} - T_\infty)}{\mu \, k \, H} \tag{3-34}$$

$Nu_b$ = *Nusselt Number (dimensionless)*

$Ra'$ = *Modified Rayleigh Number (dimensionless)*

$h_f$ = *Convection Coefficient Surrounding Fin*

$b$ = *Fin Spacing*

$k$ = *Fluid Thermal Conductivity*

$H$ = *Fin Height*

$g$ = *Acceleration Due to Gravity*

$\beta$ = *Volume Coefficient of Expansion (constant pressure)* $\left( \frac{1}{V} \left( \frac{\partial V}{\partial T} \right)_p \right)$

$T_{chas}$ = *Chassis Temperature*

$T_\infty$ = *Ambient Temperature*

$\rho$ = *Fluid Density*

$\mu$ = *Absolute Viscosity*

$c_p$ = *Fluid Specific Heat*

The fin efficiency and spacing equations may be used to optimize the fin design based upon various constraints. Some applications may require optimization based upon minimum weight for a given heat-transfer requirement. Optimization may be based upon cost or other considerations. Use of a spreadsheet (see 2.2.3) may be useful for automating the optimization process.

Underlying Assumptions

- If average bulk fluid temperature rise is used, the bulk fluid rise is small compared to the temperature rise due to convection

- The convection coefficient in a forced-convection heat exchanger is constant

- Forced-convection flow is fully turbulent

- Laminar free-convection flow [12], [13]

- Fin-spacing relationships assume laminar flow, negligible choking at top and bottom of fin, and two-dimensional flow condition [13]

**Radiation Boundary Conditions**

Background

Radiation may be a significant part of the heat transfer from an electronic enclosure, particularly if the enclosure is significantly hotter than the ambient. A relatively hot enclosure may be encountered in high power applications, or if coolant is lost in a normally forced-convection cooled configuration. Since radiation heat transfer is "line of sight", the heat exchange between two bodies is significantly influenced by the degree to which the two bodies are exposed to or "see" each other.

Radiation from Surfaces without Fins

For radiation from a body in which the surfaces are exposed only to the ambient and not to each other (such as from a box) in the radiation heat transfer is given as follows:

$$q_{rad} = \frac{\sigma \left(T_{chas}^{\,4} - T_x^{\,4}\right)}{R_{chas} + R_{if} + R_x} \tag{3-35}$$

$$R_{chas} = \frac{(1 - \epsilon_c)}{\epsilon_c \, A_{chas}} \tag{3-36}$$

$$R_{if} = \frac{1}{A_{chas}} \tag{3-37}$$

$$R_x = \frac{(1 - \epsilon_x)}{\epsilon_x \, A_x} \tag{3-38}$$

$q_{rad}$ = Radiation Heat Transfer

$\sigma$ = Stefan-Boltzmann Constant $\left(5.669x10^{-8}\dfrac{W}{m^2K^4}\right)$

$T_{chas}$ = Chassis Temperature (absolute)

$T_x$ = Ambient Temperature (absolute)

$R_{chas}$ = Chassis Radiation Surface Resistance

$R_{if}$ = Chassis to Ambient Radiation Resistance

$R_x$ = Ambient Radiation Surface Resistance

$\epsilon_c, \epsilon_x$ = Emissivity of Chassis and Ambient

$A_{chas}$ = Chassis Radiation Surface Area

If the ambient is infinitely large or is a perfect black body then $R_\infty$ becomes zero and equation 3-35 is simplified as follows

$$q_{rad} = \epsilon_c \, A_{chas} \, \sigma \, (T_{chas}^{\,4} - T_x^{\,4}) \tag{3-39}$$

Radiation from Surfaces with Fins

If parts of the electronic chassis are shielded from the ambient (such as by fins or protrusions) a radiation shape factor must be considered. This shape factor considers the amount of the chassis surfaces than can be "seen" by each other and the ambient. Since different shape factors exist for interactions between different surfaces and the ambient, an equivalent radiation resistance network is formed (see Figure 3-15) which allows the total chassis radiation to be calculated. For a typical finned chassis in a infinitely large or black ambient, heat transfer from the finned surface is given by:

$$q_{rad} = \frac{\sigma \ (T_{chas}^{\ 4} - T_{\infty}^{\ 4})}{R_{eq}} \tag{3-40}$$

$$\frac{1}{R_{eq}} = \frac{N_f}{R_{end}} + \frac{N_f - 1}{R_{space}} + \frac{2}{R_{fl}} \tag{3-41}$$

$$R_{end} = \frac{1}{\epsilon_c \ A_{end}} \tag{3-42}$$

$$A_{end} = H \ t_f \tag{3-43}$$

$q_{rad}$ = Radiation Heat Transfer from Finned Surface
$\sigma$ = Stefan-Boltzmann Constant $\left(5.669x10^{-8} \dfrac{W}{m^2 K^4}\right)$
$T_{chas}$ = Chassis Temperature (absolute)
$T_{\infty}$ = Ambient Temperature (absolute)
$R_{eq}$ = Chassis to Ambient Equivalent Resistance
$R_{end}$ = Resistance from End of Fin to Ambient
$R_{space}$ = Resistance from Space Between Fins to Ambient
$R_{fl}$ = Resistance from First/Last Fin Surface to Ambient
$N_f$ = Number of Fins
$\epsilon_c$ = Emissivity of Chassis
$A_{end}$ = Area of End of Single Fin
$H$ = Fin Height
$t_f$ = Fin Thickness

Calculation of the resistance network (see Figure 3-15) considers the shape factors between the fin space and the fin (see Table 3-3), and between adjacent fins (see Table 3-4). Numerical formulas for these shape factors can be found in the GE Heat Transfer Data Book [13] or may be derived. Equivalent radiation resistance for the space between the fins or the space from the first or last fin to the end of the chassis is given as follows:

$$R_{space},\ R_{fl} = \frac{\begin{array}{c}R_1R_3R_4 + R_1R_5R_4 + R_1R_3R_5 + R_2R_3R_4 \\ + R_2R_5R_4 + R_2R_3R_5 + R_2R_1R_4 + R_2R_1R_5\end{array}}{\begin{array}{c}R_1R_3 + R_2R_4 + R_2R_3 + R_2R_1 + \\ R_3R_4 + R_5R_4 + R_5R_3 + R_1R_5\end{array}} \qquad (3\text{-}44)$$

$$R_1 = \frac{(1 - \epsilon_c)}{\epsilon_c\, A_{space}} \qquad (3\text{-}45)$$

$$R_2 = \frac{1}{A_{space}(1 - KF_{1\text{-}2})} \qquad (3\text{-}46)$$

$$R_3 = \frac{1}{KA_{space}\, F_{1\text{-}2}} \qquad (3\text{-}47)$$

$$R_4 = \frac{(1 - \epsilon_c)}{K\epsilon_c\, A_{fin}} \qquad (3\text{-}48)$$

$$R_5 = \frac{1}{KA_{fin}[1 - F_{2\text{-}1} - (K - 1)F_{2\text{-}2}]} \qquad (3\text{-}49)$$

$$A_{space} = H\, b \qquad (3\text{-}50)$$

$$A_{fin} = H\, L_f \qquad (3\text{-}51)$$

$$F_{2\text{-}1} = \frac{A_{space}}{A_{fin}} F_{1\text{-}2} \qquad (3\text{-}52)$$

$R_1, R_2, R_3, R_4, R_5$ = *Radiation Resistances (see Figure 3-15)*

$A_{space}$ = *Area of Space between Fins, Area from First/Last Fin to End*

$K$ = *1 for First/Last Fin, 2 for Space Between Fins*

$F_{1\text{-}2}$ = *View Factor from Fin Space to Fin (see Table 3-3)*

$F_{2\text{-}1}$ = *View Factor from Fin to Fin Space*

$F_{2\text{-}2}$ = *View Factor from Fin to Fin  (see Table 3-4)*

$A_{fin}$ = *Area of Face of Fin*

$H$ = *Fin Height*

$b$ = *Fin Spacing, Space from First/Last Fin to End of Chassis*

$L_f$ = *Actual Fin Length*


The fin-to-fin shape factor ($F_{2\text{-}2}$) in equation 3-49 is not required for the first/last fin since the ($K$-1) term goes to zero. Physically this makes sense since the outer faces of the first/last fin only sees the space between the fin and the end of the chassis and does not see any other fins. In the case of the equivalent network for the space between the fins $R_3$, $R_4$, and $R_5$ (equations 3-47 through 3-49) are divided by two from that shown in Figure 3-15 to represent the effect of the equal resistances in parallel.

Underlying Assumptions

- Fin temperature for radiation purposes is constant

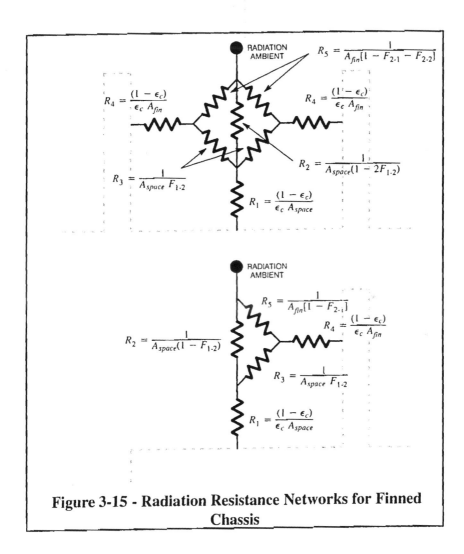

**Figure 3-15 - Radiation Resistance Networks for Finned Chassis**

| Fin Space*/ Height (b/H) | Fin Length/Height ($L_f$/H) | | | | | |
|---|---|---|---|---|---|---|
| | **0.05** | **0.10** | **0.15** | **0.20** | **0.25** | **0.30** |
| **0.03** | 0.3563 | 0.4184 | 0.4404 | 0.4513 | 0.4576 | 0.4617 |
| **0.04** | 0.3192 | 0.3945 | 0.4228 | 0.4370 | 0.4454 | 0.4509 |
| **0.05** | 0.2874 | 0.3720 | 0.4060 | 0.4234 | 0.4337 | 0.4404 |
| **0.06** | 0.2603 | 0.3511 | 0.3899 | 0.4102 | 0.4224 | 0.4304 |
| **0.08** | 0.2176 | 0.3137 | 0.3598 | 0.3853 | 0.4009 | 0.4112 |
| **0.10** | 0.1860 | 0.2819 | 0.3327 | 0.3622 | 0.3807 | 0.3932 |
| **0.12** | 0.1620 | 0.2549 | 0.3083 | 0.3408 | 0.3618 | 0.3762 |
| **0.16** | 0.1282 | 0.2124 | 0.2669 | 0.3030 | 0.3275 | 0.3450 |
| **0.20** | 0.1058 | 0.1811 | 0.2339 | 0.2710 | 0.2977 | 0.3171 |
| **0.20** | 0.1058 | 0.1811 | 0.2339 | 0.2710 | 0.2977 | 0.3171 |

**Table 3-3 - Shape Factors for Space Between Fins to Fin ($F_{1-2}$)**

(see Figure 3-14 for dimensions)

\* - Spacing between fins or space from fin to end of chassis

**Chassis Thermal Mass**

<u>Background</u>

Reiterating that thermal masses need only be considered in cases where transient thermal analysis is required (see 3.3.2), the thermal mass for the chassis is determined by taking the product of the chassis mass and the appropriate specific heat.

$$Mc_{chas} = m_c \, c_c \qquad (3\text{-}53)$$

$Mc_{chas}$ = *Chassis Thermal Mass*
$m_c$ = *Mass of Chassis*
$c_c$ = *Specific Heat of Chassis*

If considerable temperature variation in the chassis is present, the distribution of thermal mass may need to be considered along with the thermal resistance distribution within the chassis. Finite-element analysis (see 3.6.3) is typically used for such situations.

Underlying Assumptions

- Chassis is at a uniform average temperature

## 3.6.3   Finite-Element Analysis

In cases where the chassis is not at a uniform temperature or complex thermal paths exist, a finite-element analysis is typically required (see 3-16). A non-uniform chassis temperature distribution typically occurs when the chassis wall sections are thin or the material conductivity is relatively low. Thermal finite-element models typically include only one degree-of-freedom per node (see 3.4.2) but since the radiation and free-convection boundary conditions are non-linear, an iterative solution is typically required. Although a chassis finite-element thermal model requires more calculation time due to the iterative solution than a component or module model, the solution time is usually less than a comparable structural model. Finite-element analysis is especially appropriate if an electronic method is available for describing the chassis geometry such as an IGES file.

| Table 3-4 - Shape Factors for Fin to Fin ($F_{2-2}$) | | | | | | |
|---|---|---|---|---|---|---|
| (see Figure 3-14 for dimensions) | | | | | | |
| Fin Length/ Space ($L_f/b$) | Fin Height/Space (H/b) | | | | | |
| | 4 | 6 | 9 | 13 | 20 | 30 |
| 0.5 | 0.1986 | 0.2108 | 0.2191 | 0.2243 | 0.2284 | 0.2310 |
| 0.75 | 0.2795 | 0.2970 | 0.3090 | 0.3164 | 0.3223 | 0.3260 |
| 1 | 0.3460 | 0.3681 | 0.3833 | 0.3928 | 0.4003 | 0.4049 |
| 1.5 | 0.4437 | 0.4733 | 0.4937 | 0.5064 | 0.5164 | 0.5227 |
| 2 | 0.5090 | 0.5442 | 0.5685 | 0.5836 | 0.5956 | 0.6031 |
| 2.5 | 0.5545 | 0.5940 | 0.6212 | 0.6383 | 0.6518 | 0.6602 |
| 3.5 | 0.6126 | 0.6580 | 0.6895 | 0.7093 | 0.7250 | 0.7347 |
| 5 | 0.6602 | 0.7109 | 0.7464 | 0.7687 | 0.7865 | 0.7976 |
| 7 | 0.6936 | 0.7484 | 0.7869 | 0.8113 | 0.8308 | 0.8429 |
| 10 | 0.7193 | 0.7774 | 0.8186 | 0.8447 | 0.8656 | 0.8787 |

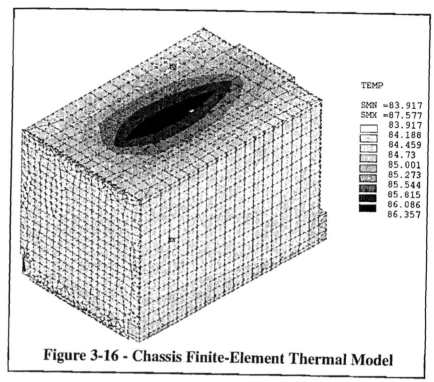

TEMP

SMN =83.917
SMX =87.577

- [ ] 83.917
- [ ] 84.188
- [ ] 84.459
- [ ] 84.73
- [ ] 85.001
- [ ] 85.273
- [ ] 85.544
- [ ] 85.815
- [ ] 86.086
- [ ] 86.357

## Figure 3-16 - Chassis Finite-Element Thermal Model

# 3.7   COMBINING ALL PACKAGING LEVELS

The thermal performance at each packaging level can be combined to determine the overall system thermal performance. This uses the thermal performance at chassis, module, and component levels to determine the overall system thermal performance for a given environment.

In some cases the effect of the chassis thermal management on the ambient needs to be considered. This is especially true when a electronic enclosure is installed in a confined space that does not allow the full free-convection conditions to occur. In such cases, the heating of the ambient by the electronic equipment installed in the space needs to be considered. Also significant is the effect of radiation from adjacent electronic equipment. Readers are encouraged to develop their own relationships for these cases.

## 3.7.1   Steady-State Thermal Analysis

Steady-state thermal performance is determined by first determining the chassis temperature. Since the free-convection and radiation relationships are non-linear, direct calculation of the chassis temperature is not practical. Steady-state chassis temperatures may be determined by assuming a chassis temperature and calculating the heat loss considering forced convection, free convection, and radiation heat loss as appropriate. If the the heat loss does not match the total internal power, the assumed temperature may be iterated to determine the chassis temperature. Spreadsheet and symbolic math programs often include methods for solving non-linear iterative relationships.

The module temperatures can be determined by using the local chassis temperature rise near the module (if applicable), the chassis to module interface resistance, and the thermal resistance of the module itself to determine the overall thermal resistance of the modules. This thermal resistance can be combined with the module power and chassis temperature to determine the module temperature.

Component temperatures are found by using the component thermal resistance in conjunction with the component power and module temperature to determine the component junction temperature. Once the component junction temperature is determined it can be compared with component ratings to determine survivability in the specified environment.

## 3.7.2    Transient Thermal Analysis

Transient thermal analysis requires calculating a temperature of the chassis, and each of the modules as a function of time. Transient thermal performance may be expressed using differential equations, as follows:

$$Mc_{chas}\frac{dT_{chas}}{dt} = P_{tot} - \Sigma\frac{(T_i - T_{chas})}{\theta_i} - q_{rad} - q_{free} - q_{forced} \qquad (3\text{-}54)$$

$$Mc_{mod}\frac{dT_i}{dt} = \frac{(T_{chas} - T_i)}{\theta_i} + P_i \qquad (3\text{-}55)$$

$Mc_{chas}$ = Chassis Thermal Mass

$T_{chas}$ = Average Chassis Temperature

$t$ = Time

$P_{tot}$ = Total Power Dissipated within Chassis

$T_i$ = Temperature of Module i

$\theta_i$ = Thermal Resistance of Module i (average module to chassis, see text)

$q_{rad}$ = Heat Loss due to Radiation

$q_{free}$ = Heat Loss due Free Convection

$q_{forced}$ = Heat Loss due to Forced Convection

$Mc_i$ = Thermal Mass of Module i

$T_i$ = *Temperature of Module i*

$P_i$ = *Power of Module i*

The thermal resistance of the module represents thermal resistance derived by taking the steady-state difference between average module heat sink temperature and the average chassis, and dividing by the module power. This thermal resistance includes the temperature rise of the module heat sink, the chassis to module interface, and the local temperature rise of the the chassis (where applicable).

Since the free-convection and radiation heat loss rates are non-linear, these differential equations are non-linear so a numerical solution is typically required. The numerical solution can be coded in a spreadsheet (see 2.2.3), a custom program (see 2.2.5), or simplified finite-element analysis using only lumped masses and resistances.

### 3.7.3    Finite-Element Analysis Guidelines

The following finite-element analysis guidelines are applicable to all levels of thermal performance analysis:

**Element Density**

It is important that the mesh density allows the temperature profile to be properly defined. If an increased mesh density is not practical, the use of higher order elements with mid side nodes (see Figure 4-2) may be used to improve accuracy. When higher order elements are used, it is important to avoid connecting the midside nodes of one element to the corners of another.

**Meshing guidelines**

Typically a finer mesh is required in the area of expected heat flux concentrations. If accurate results are not required in the area of a heat concentration, a coarser mesh may be adequate. Alternatively, the cut boundary method may be used to determine the temperature profile near a heat concentration. In the cut boundary method, a relatively coarse mesh is used to determine the overall temperatures across a part, these temperatures are then used as boundary conditions on a finer mesh model which models the area of the heat concentration.

**Other Inconsistencies**

Occasionally a model may have inconsistencies (usually due to the modeling method) which may not be apparent with geometry plotting. Some situations to watch for include:

- Nodes at the same location but connected to different elements

- Boundary conditions do not match actual situation

- Boundary conditions between different levels of the analysis are not consistent

- Material types improperly assigned

- Total mass of model does not match total system mass

## 3.7.4    Typical Thermal Calculation

Consider an aluminum (6061) electronic chassis painted with black gloss lacquer with modules as shown in Figure 3-17. Each of the five surface-mount technology modules (aluminum heat sink) dissipates 25 Watts for a total of 125 Watts. PWBs are bonded to the heat sink with 10 mils (0.010 in) of silicone rubber.  For operation in a 25°C ambient, determine the following steady-state temperatures:

- Average chassis temperature

- Average module temperature

- Junction temperature for a hypothetical surface-mount component with the following details:

    - Power dissipation = 1 Watt (included in 25-Watt module total)
    - Package footprint dimensions = 0.95 in x 0.95 in
    - Junction-to-case temperature rise = 5°C/W
    - No thermal vias under component

**Average Chassis Temperature**

The steady-state chassis temperature is determined by first assuming a chassis temperature and calculating the heat loss considering free-convection, and radiation heat loss as appropriate.  Assuming an initial chassis temperature of 50°C, the convection and radiation heat loss is determined as follows:

Convection Coefficients for Surfaces without Fins

For the surfaces without fins (top, bottom, front, and back):

$T_{chas}$ = Chassis Temperature = 50°C

$T_{\infty}$ = Ambient Temperature = 25°C

$g$ = Acceleration Due to Gravity = 386.4 $\frac{in}{sec^2}$

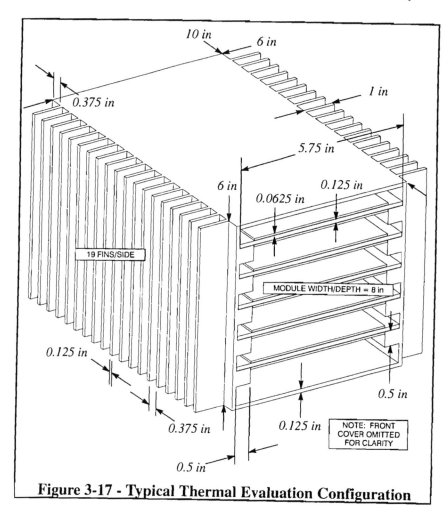

**Figure 3-17 - Typical Thermal Evaluation Configuration**

Ambient air properties ([12], [16]) evaluated at film temperature of 37.5°C:

$$k = Fluid\ Thermal\ Conductivity = 6.90\ x\ 10^{-4}\frac{W}{in\text{-}°C}$$

$$\beta = Volume\ Coefficient\ of\ Expansion = 3.22\ x\ 10^{-3}\frac{1}{°C}$$

$$\rho = Fluid\ Density = 4.10\ x\ 10^{-5}\frac{lbm}{in^3}$$

$$\mu = Absolute\ Viscosity = 1.06x10^{-6}\frac{lbm}{sec\text{-}in}$$

$$c_p = Fluid\ Specific\ Heat = 455.8\ \frac{J}{lbm\text{-}°C}$$

$$Pr = Prandtl\ Number = \frac{c_p\mu}{k} = 0.700$$

For the front and back (ends):

$$a = Characteristic\ Dimension = Height = 6\ in$$

$$Gr_a = \frac{g\beta(T_{chas} - T_\infty)a^3\rho^2}{\mu^2}$$

$$Gr_a = \frac{386.4\frac{in}{sec^2}3.22\ x\ 10^{-3}\frac{1}{°C}(50°C - 25°C)(6\ in)^3(4.10\ x\ 10^{-5}\frac{lb}{in^3})^2}{(1.06x10^{-6}\frac{lb}{sec\text{-}in})^2} \quad (3\text{-}26)$$

$$Gr_a = 1.01x10^7$$

From Table 3-2, C = 0.59 and m = 0.25:

$$Nu_a = C(Gr_a\ Pr)^m = 0.59\ (\ 1.01x10^7\ x\ 0.700\ )^{0.25} = 30.4 \quad (3\text{-}25)$$

$$h_e = h = \frac{Nu_a\ k}{a} = \frac{30.4\ x\ 6.90\ x\ 10^{-4}\frac{W}{in\text{-}°C}}{6\ in} = 3.50x10^{-3}\frac{W}{in^2°C} \quad (3\text{-}24)$$

For the top:

$$a = Characteristic\ Dimension = Mean\ of\ Sides = \frac{(6\ in\ +\ 10\ in)}{2} = 8\ in$$

$$Gr_a = \frac{386.4\frac{in}{sec^2}3.22\ x\ 10^{-3}\frac{1}{°C}(50°C - 25°C)(8\ in)^3(4.10\ x\ 10^{-5}\frac{lb}{in^3})^2}{(1.06x10^{-6}\frac{lb}{sec\text{-}in})^2} \quad (3\text{-}26)$$

$$Gr_a = 2.38x10^7$$

From Table 3-2, C = 0.14 and m = 0.333:

$$Nu_a = C(Gr_a\ Pr)^m = 0.14\ (\ 2.38x10^7\ x\ 0.700\ )^{0.333} = 35.6 \quad (3\text{-}25)$$

$$h_t = h = \frac{Nu_a\ k}{a} = \frac{35.6\ x\ 6.90\ x\ 10^{-4}\frac{W}{in\text{-}°C}}{8\ in} = 3.07x10^{-3}\frac{W}{in^2°C} \quad (3\text{-}24)$$

For the bottom:

$$a = Characteristic\ Dimension = Mean\ of\ Sides = \frac{(6\ in\ +\ 10\ in)}{2} = 8\ in$$

$$Gr_a = \frac{386.4\frac{in}{sec^2}3.22\ x\ 10^{-3}\frac{1}{°C}(50°C - 25°C)(8\ in)^3(4.10\ x\ 10^{-5}\frac{lb}{in^3})^2}{(1.06x10^{-6}\frac{lb}{sec\text{-}in})^2} \tag{3-26}$$

$$Gr_a = 2.38x10^7$$

From Table 3-2, C = 0.27 and m = 0.25:

$$Nu_a = C(Gr_a\ Pr)^m = 0.27\ (\ 2.38x10^7\ x\ 0.700\ )^{0.25} = 17.2 \tag{3-25}$$

$$h_b = h = \frac{Nu_a\ k}{a} = \frac{17.2\ x\ 6.90\ x\ 10^{-4}\frac{W}{in\text{-}°C}}{8\ in} = \mathbf{1.48x10^{-3}}\frac{W}{in^2°C} \tag{3-24}$$

### Convection Coefficient for Surface with Fins

$$b = Fin\ Spacing = 0.375in$$
$$H = Fin\ Height = 6\ in$$

$$Ra' = \frac{(0.000041\frac{lb}{in^3})^2 455.8\frac{J}{lb\text{-}°C}386.4\frac{in}{sec^2}0.00322\frac{1}{°C}(0.375\ in)^4(50°C - 25°C)}{1.06x10^{-6}\frac{lb}{sec\text{-}in}\ 6.90\ x\ 10^{-4}\frac{W}{in\text{-}°C}\ 6\ in} \tag{3-34}$$

$$Ra' = 107$$

$$Nu_b = \left[\frac{576}{(Ra')^2} + \frac{2.87}{\sqrt{Ra'}}\right]^{-\frac{1}{2}} = \left[\frac{576}{(107)^2} + \frac{2.87}{\sqrt{107}}\right]^{-\frac{1}{2}} = 1.75 \tag{3-33}$$

$$h_f = \frac{Nu_b\ k}{b} = \frac{1.75\ x\ 6.90\ x\ 10^{-4}\frac{W}{in\text{-}°C}}{0.375\ in} = \mathbf{3.22x10^{-3}}\frac{W}{in^2°C} \tag{3-32}$$

### Fin Efficiency

$$H = Fin\ Height = 6\ in$$
$$t_f = Fin\ Thickness = 0.125\ in$$
$$h_f = Convection\ Coefficient\ Surrounding\ Fin = 3.22x10^{-3}\frac{W}{in^2°C}$$
$$L_f = Actual\ Fin\ Length = 1\ in$$

For 6061 Aluminum [17]:

$$k_f = Fin\ Thermal\ Conductivity = 104\frac{Btu}{hr\text{-}ft°F} = 4.57\frac{W}{in\text{-}°C}$$

$$A_f = H\ t_f = 6\ in\ x\ 0.125\ in = 0.75\ in^2 \tag{3-31}$$

$$P_f = 2(H + t_f) = 2(6\ in + 0.125\ in) = 12.25\ in \tag{3-30}$$

$$m = \sqrt{\frac{h_f\ P_f}{k_f\ A_f}} = \sqrt{\frac{3.22x10^{-3}\frac{W}{in^2°C}\ 12.25\ in}{4.57\frac{W}{in\text{-}°C}\ 0.75\ in^2}} = 0.107\frac{1}{in} \tag{3-29}$$

$$L_c = L_f + \frac{t_f}{2} = 1\ in + \frac{0.125\ in}{2} = 1.0625\ in \tag{3-28}$$

$$\eta_f = \frac{\tanh(mL_c)}{mL_c} = \frac{\tanh(0.107\frac{1}{in}\ x\ 1.0625\ in)}{0.107\frac{1}{in}\ x\ 1.0625\ in} = \mathbf{0.9957} \tag{3-27}$$

## Overall Convection Heat Transfer

$\eta_f = Fin\ Efficiency = 0.9957$

$h_s = h_f = Convection\ Coefficient\ on\ Finned\ Sides = 3.22x10^{-3}\frac{W}{in^2°C}$

$h_t = Convection\ Coefficient\ on\ Top = 3.07x10^{-3}\frac{W}{in^2°C}$

$h_b = Convection\ Coefficient\ on\ Bottom = 1.48x10^{-3}\frac{W}{in^2°C}$

$h_e = Convection\ Coefficient\ on\ Ends = 3.50x10^{-3}\frac{W}{in^2°C}$

$A_s = Area\ of\ Sides\ (excluding\ fins) = 2(10in - 19fins\ x\ 0.125in/fin)\ x\ 6in$

$A_s = 91.5in^2$

$A_t = A_b = Area\ of\ Top,\ Bottom = 10\ in\ x\ 6\ in = 60\ in^2$

$A_e = Area\ of\ Ends = 2(6\ in\ x\ 6\ in) = 72\ in^2$

$A_f = Surface\ Area\ of\ Fins = 2\ x\ 19\ x\ L_c P_f$

$$P_f = 12.25\ in \tag{3-30}$$

$$L_c = 1.0625\ in \tag{3-28}$$

$A_f = 2\ x\ 19\ x\ (1.0625\ in\ x\ 12.25\ in) = 494.6\ in^2$

$$q_{free} = (3.22x10^{-3}\frac{W}{in^2\,^\circ C}91.5in^2 + 3.07x10^{-3}\frac{W}{in^2\,^\circ C}60\,in^2 +$$

$$1.48x10^{-3}\frac{W}{in^2\,^\circ C}60\,in^2 + 3.50x10^{-3}\frac{W}{in^2\,^\circ C}72\,in^2 + \qquad (3\text{-}23)$$

$$+ 0.9957x3.22x10^{-3}\frac{W}{in^2\,^\circ C}494.6\,in^2)\,(50^\circ C - 25^\circ C)$$

$$q_{free} = \mathbf{60.1\ W}$$

### Radiation from Surfaces without Fins

For shiny black lacquer coating [18] evaluated at a temperature of 37.5°C

$\epsilon_c$ = *Emissivity of Chassis* = 0.87

$A_{chas}$ = *Chassis (without fins) Radiation Surface Area*

$A_{chas} = A_t + A_b + A_e = 60\,in^2 + 60\,in^2 + 72\,in^2 = 192\,in^2$

$\sigma$ = *Stefan Boltzmann Constant* = $3.657x10^{-11}\frac{W}{in^2K^4}$

$T_{chas}$ = *Chassis Temperature* = $50^\circ C = 323^\circ K$

$T_\infty$ = *Ambient Temperature* = $25^\circ C = 298^\circ K$

$$q_{rad} = \epsilon_c\,A_{chas}\,\sigma\,(T_{chas}^4 - T_\infty^4)$$

$$q_{rad} = 0.87x192\,in^2\,3.657x10^{-11}\frac{W}{in^2K^4}\,((323^\circ K)^4 - (298^\circ K)^4)\qquad(3\text{-}39)$$

$$q_{rad} = \mathbf{18.3\ W}$$

### Radiation from Surfaces with Fins

For space between fins:

$\epsilon_c$ = *Emissivity of Chassis* = 0.87

$H$ = *Fin Height* = 6 in

$b$ = *Fin Spacing* = 0.375 in

$L_f$ = *Actual Fin Length* = 1 in

$K$ = 2 *for Space Between Fins*

(interpolating from Table 3-3, b/H = 0.0625 and $L_f$/H = 0.1667)

$F_{1\text{-}2}$ = *View Factor from Fin Space to Fin* = 0.3931

(interpolating from Table 3-4, H/b = 16 and $L_f$/b = 2.6667)

$F_{2\text{-}2}$ = *View Factor from Fin to Fin* = 0.6561

$$A_{space} = H\,b = 6\ in \times 0.375\ in = 2.25\ in^2 \tag{3-50}$$

$$R_1 = \frac{(1 - \epsilon_c)}{\epsilon_c\,A_{space}} = \frac{(1 - 0.87)}{0.87x2.25\ in^2} = 0.06641\frac{1}{in^2} \tag{3-45}$$

$$R_2 = \frac{1}{A_{space}(1 - KF_{1-2})} = \frac{1}{2.25in^2(1 - 2x0.3931)} = 2.0788\frac{1}{in^2} \tag{3-46}$$

$$R_3 = \frac{1}{KA_{space}\,F_{1-2}} = \frac{1}{2x2.25in^2x0.3931} = 0.5653\frac{1}{in^2} \tag{3-47}$$

$$A_{fin} = H\,L_f = 6\ in \times 1\ in = 6\ in^2 \tag{3-51}$$

$$R_4 = \frac{(1 - \epsilon_c)}{K\epsilon_c\,A_{fin}} = \frac{(1 - 0.87)}{2x0.87x6in^2} = 0.01245\frac{1}{in^2} \tag{3-48}$$

$$F_{2-1} = \frac{A_{space}}{A_{fin}}F_{1-2} = \frac{2.25\ in^2}{6\ in^2}0.3931 = 0.1474 \tag{3-52}$$

$$R_5 = \frac{1}{KA_{fin}[1 - F_{2-1} - (K - 1)F_{2-2}]} = \frac{1}{2x6in^2(1 - 0.1474 - 0.6561)}$$
$$\tag{3-49}$$
$$R_5 = 0.4241\frac{1}{in^2}$$

$$R_{space} = \frac{\begin{array}{c}R_1R_3R_4 + R_1R_5R_4 + R_1R_3R_5 + R_2R_3R_4 \\ + R_2R_5R_4 + R_2R_3R_5 + R_2R_1R_4 + R_2R_1R_5\end{array}}{\begin{array}{c}R_1R_3 + R_2R_4 + R_2R_3 + R_2R_1 + \\ R_3R_4 + R_5R_4 + R_5R_3 + R_1R_5\end{array}} \tag{3-44}$$

$$R_{space} = \frac{0.6010\frac{1}{in^6}}{1.6568\frac{1}{in^4}} = 0.3627\frac{1}{in^2}$$

For first/last fins:

$\epsilon_c = Emissivity\ of\ Chassis = 0.87$

$H = Fin\ Height = 6\ in$

$b = Space\ from\ First/Last\ Fin\ to\ End\ of\ Chassis = 0.375\ in$

$L_f = Actual\ Fin\ Length = 1\ in$

$K = 1\ for\ First/Last\ Fin$

(interpolating from Table 3-3, b/H = 0.0625 and $L_f$/H = 0.1667)

$$F_{1-2} = View\ Factor\ from\ Fin\ Space\ to\ Fin = 0.3931$$

$$A_{space} = H\,b = 6\ in \times 0.375\ in = 2.25\ in^2 \tag{3-50}$$

$$R_1 = \frac{(1 - \epsilon_c)}{\epsilon_c \, A_{space}} = \frac{(1 - 0.87)}{0.87x2.25 \, in^2} = 0.06641 \frac{1}{in^2} \tag{3-45}$$

$$R_2 = \frac{1}{A_{space}(1 - KF_{1-2})} = \frac{1}{2.25in^2(1 - 0.3931)} = 0.7323 \frac{1}{in^2} \tag{3-46}$$

$$R_3 = \frac{1}{KA_{space} \, F_{1-2}} = \frac{1}{2.25in^2x0.3931} = 1.1306 \frac{1}{in^2} \tag{3-47}$$

$$A_{fin} = H \, L_f = 6 \, in \, x \, 1 \, in = 6 \, in^2 \tag{3-51}$$

$$R_4 = \frac{(1 - \epsilon_c)}{K\epsilon_c \, A_{fin}} = \frac{(1 - 0.87)}{0.87x6in^2} = 0.02490 \frac{1}{in^2} \tag{3-48}$$

$$F_{2-1} = \frac{A_{space}}{A_{fin}}F_{1-2} = \frac{2.25 \, in^2}{6 \, in^2}0.3931 = 0.1474 \tag{3-52}$$

$$R_5 = \frac{1}{A_{fin}(1 - F_{2-1})} = \frac{1}{6in^2(1 - 0.1474)} = 0.1955 \frac{1}{in^2} \tag{3-49}$$

$$R_{fl} = \frac{\begin{array}{c} R_1R_3R_4 + R_1R_5R_4 + R_1R_3R_5 + R_2R_3R_4 \\ + \, R_2R_5R_4 + R_2R_3R_5 + R_2R_1R_4 + R_2R_1R_5 \end{array}}{\begin{array}{c} R_1R_3 + R_2R_4 + R_2R_3 + R_2R_1 + \\ R_3R_4 + R_5R_4 + R_2R_3 + R_1R_5 \end{array}} \tag{3-44}$$

$$R_{fl} = \frac{0.2136 \frac{1}{in^6}}{1.2369 \frac{1}{in^4}} = 0.1727 \frac{1}{in^2}$$

Total radiation heat transfer:

$H = Fin \, Height = 6 \, in$

$t_f = Fin \, Thickness = 0.125 \, in$

$\epsilon_c = Emissivity \, of \, Chassis = 0.87$

$N_f = Number \, of \, Fins = 19$

$R_{fl} = Resistance \, from \, First/Last \, Fin \, Surface \, to \, Ambient = 0.1727 \frac{1}{in^2}$

$R_{space} = Resistance \, from \, Space \, Between \, Fins \, to \, Ambient = 0.3627 \frac{1}{in^2}$

$\sigma = Stefan \, Boltzmann \, Constant \, \left(3.657x10^{-11}\frac{W}{in^2K^4}\right)$

$T_{chas} = Chassis \, Temperature = 50°C = 323°K$

$T_\infty = Ambient \, Temperature = 25°C = 298°K$

$$A_{end} = H \, t_f = 6 \, in \, x \, 0.125 \, in = 0.75 \, in^2 \qquad (3\text{-}43)$$

$$R_{end} = \frac{1}{\epsilon_c \, A_{end}} = \frac{1}{0.87x0.75in^2} = 1.5326\frac{1}{in^2} \qquad (3\text{-}42)$$

$$\frac{1}{R_{eq}} = \frac{N_f}{R_{end}} + \frac{N_f - 1}{R_{space}} + \frac{2}{R_{fl}}$$

$$\frac{1}{R_{eq}} = \frac{19}{1.5326\frac{1}{in^2}} + \frac{19 - 1}{0.3627\frac{1}{in^2}} + \frac{2}{0.1727\frac{1}{in^2}} = 73.606 \, in^2$$

$$(3\text{-}41)$$

$$q_{rad} = \frac{\sigma \, (T_{chas}^{\,4} - T_\infty^{\,4})}{R_{eq}}$$

$$(3\text{-}40)$$

$$q_{rad} = 3.657x10^{-11}\frac{W}{in^2K^4} \, ((323°K)^4 - (298°K)^4)x73.606in^2$$

$$q_{rad} = \mathbf{8.1 \, W} \, (per \, side)$$

## Chassis Temperature Calculation

$$Total \; Heat \; Transfer = 2 \; sides \; x \; 8.1\frac{W}{side} + 18.3W + 60.2W = \mathbf{94.7 \; W}$$

Since the total power transferred to the environment of 94.7 Watts is less than the 125-Watt total power dissipation, the initial chassis estimate of 50°C must be increased until the power transferred to the environment equals the 125-Watt internal dissipation. One method of solving this problem is the use of the bisection method. In this method, the desired solution is bracketed by a temperature range and the range is split in half with each subsequent iteration until the temperature range or heat loss is within some tolerance. This method converges quite rapidly to the desired solution:

| Chassis Temperature (°C) | Heat Loss (W) |
|---|---|
| 50 | 94.7 |
| 60 | 144.7 |
| 55 | 119.1 |
| 57.5 | 131.7 |
| 56.25 | 125.4 |
| 55.625 | 122.2 |
| 55.9375 | 123.8 |
| 56.09375 | 124.6 |
| **56.171875** | 125.0 |

So the *average chassis temperature* for the conditions described above is *56.2°C*. For simplicity, material properties at 37.5°C were used for these calculations which is relatively close to the actual 40.6°C film temperature. In an actual analysis, the effect of film temperature variation on material properties should be considered.

**Average Module Temperature**

<u>Local Chassis Temperature Rise</u>

$w$ = *Width of Chassis Rail (module length)* = 8 *in*
$t$ = *Thickness of Chassis Rail* = 0.5 *in*
$l$ = *Length (along heat flow path) of Chassis Rail* = 0.5 *in*

For 6061 Aluminum [17]:

$$k = \textit{Thermal Conductivity} = 104 \frac{Btu}{hr\text{-}ft°F} = 4.57 \frac{W}{in\text{-}°C}$$

$$\theta = \frac{l}{k\ w\ 2t} = \frac{0.5\ in}{4.57\frac{W}{in\text{-}°C}x8inx2x0.5in} = 0.0137\frac{°C}{W} \tag{3-13}$$

$$\textit{Local Chassis Temperature Rise} = 25W\ x\ 0.0137\frac{°C}{W} = \mathbf{0.34°C}$$

<u>Module-to-Chassis Interface</u>

$$A_C = \textit{Contact Area} = 2\ x\ 8in\ x\ 0.5in = 8\ in^2$$

Thermal contact resistivity depends upon contact force, material and plating properties, and use of thermal enhancement materials. Actual values are best obtained from testing. For the purposes of this example, assume:

$$R_c = \textit{Thermal Contact Resistivity} = 0.5\frac{°C\text{-}in^2}{W}$$

$$\theta_c = \frac{R_c}{A_c} = \frac{0.5\frac{°C\text{-}in^2}{W}}{8in^2} = 0.0625\frac{°C}{W} \tag{3-7}$$

$$\textit{Module to Chassis Interface Temperature Rise} = 25Wx0.0625\frac{°C}{W} = \mathbf{1.56°C}$$

## Surface-Mount Technology Module

$P = Module\ Power = 25\ W$

$w = Width\ (depth)\ of\ Heat\ Sink = 8\ in$

$t = Thickness\ of\ Heat\ Sink = 0.125\ in$

$l = Length/Span\ of\ Heat\ Sink = 5.75\ in$

For 6061 Aluminum [17]:

$$k_{hs} = Fin\ Thermal\ Conductivity = 104\frac{Btu}{hr\text{-}ft°F} = 4.57\frac{W}{in\text{-}°C}$$

$$\Delta T_{avg} = \frac{Pl}{12k_{hs}wt} = \frac{25Wx5.75in}{12x4.57\frac{W}{in\text{-}°C}x8inx0.125in} = 2.62°C \qquad (3\text{-}12)$$

## Average Module Temperature

The average module temperature is given by the sum of the average chassis temperature and the local, interface, and module temperature rises:

$$Average\ Module\ Temperature = 56.2°C + 0.34°C + 1.56°C + 2.62°C$$

$$\boldsymbol{Average\ Module\ Temperature = 60.7°C}$$

## Component Junction Temperature

### Thermal Resistance to Heat Sink

$t_b = Thickness\ of\ Bond = 0.010\ in$

$t_{pwb} = Thickness\ of\ Bond\ and\ PWB = 0.0625\ in$

$A_c = A_d = Area\ of\ Component,\ PWB = 0.95inx0.95in = 0.9025in^2$

$A_{cu} = Area\ of\ Copper\ (thermal\ vias) = 0$

For silicone rubber [17] and glass/epoxy [19]:

$$k_b = Thermal\ Conductivity\ of\ Bond = 0.13\frac{Btu}{hr\text{-}ft°F} = 0.0057\frac{W}{in\text{-}°C}$$

$$k_d = Thermal\ Conductivity\ of\ PWB = 0.2\frac{Btu}{hr\text{-}ft°F} = 0.0088\frac{W}{in\text{-}°C}$$

$$\theta_l = \frac{0.010in}{0.0057\frac{W}{in\text{-}°C}x0.9025in^2} + \frac{0.0625in}{0.0088\frac{W}{in\text{-}°C}x0.9025in^2 + 0} = 9.81\frac{°C}{W} \qquad (3\text{-}8)$$

### Thermal Contact between Component Body and Module

Thermal contact between the component body and module is determined by dividing a thermal contact resistivity parameter by the area of contact in order to determine the thermal resistance:

$$A_c = Area\ of\ Component = 0.95 in \times 0.95 in = 0.9025 in^2$$

Thermal contact resistivity depends upon contact force, material and plating properties, use of thermal enhancement materials. Actual values are best obtained from testing. For the purposes of this example, assume:

$$R_c = Thermal\ Contact\ Resistivity = 1.0 \frac{°C\text{-}in^2}{W}$$

$$\theta_c = \frac{R_c}{A_c} = \frac{1.0 \frac{°C\text{-}in^2}{W}}{0.9025 in^2} = 1.11 \frac{°C}{W} \qquad (3\text{-}7)$$

### Component Junction Temperature

Component junction temperature is the sum of the module temperature, and the total thermal resistance from the junction to the heat sink (junction-to-case, contact, and component to heat sink) times the component power:

$$Junction\ Temperature = 60.7°C + \left(5° \frac{C}{W} + 9.81° \frac{C}{W} + 1.11° \frac{C}{W}\right) \times 1\ W$$

$$\textbf{Junction Temperature} = \textbf{76.6°C}$$

## 3.8   SPECIAL THERMAL CASES

In some cases, the conventional heat-transfer methods described in this chapter are not adequate for high-power and/or high density packaging configurations. In such situations, methods for enhancing the thermal performance are desired. Five methods for improving thermal performance are:

- Flow-through modules
- Heat pipes
- Thermoelectric coolers
- Immersion cooling
- Phase change material

In any case where special heat transfer techniques are considered it is important to verify that the power dissipation estimates are correct to avoid adding unnecessary cost.

## 3.8.1    Flow-through Modules

A flow-through module (see Figure 3-18) incorporates the flow-through chassis cooling configuration in a (typically surface-mount) module. This eliminates the module-to-chassis thermal resistance and provides a large increase in the power handling capability of the module. Precautions must be taken to avoid leaks during operation of the module. If a liquid cooling medium is used, additional precautions may need to be taken to prevent leaks when the module is removed such as by the use of self-sealing quick disconnects. The thermal performance of configurations using flow-through modules may be predicted by using the heat transfer relationships for forced-convection cooling of a chassis in place of the module thermal performance.

**Figure 3-18 - Flow-Through Modules**
(Courtesy of Lockheed Martin Control Systems.)

## 3.8.2    Heat Pipes

A heat pipe (see Figure 3-19) consists of a sealed tube containing a condensable fluid with a wicking material on the inside surface of the tube. As heat is added to one end of the heat pipe, the fluid evaporates and travels as a gas to the other end of the heat pipe. When heat is removed from the other end, the fluid condenses and travels by capillary action along the wick back to the heated end. Thermal performance of the heat pipe depends upon the selection of fluid, tube material, wick configuration, and other design parameters as described in Chi [14]. In some cases the heat pipe may be almost isothermal up to relatively high heat transfer rates. In many cases the largest temperature rise is observed at the interface between the heat pipe and the heat source/sink.

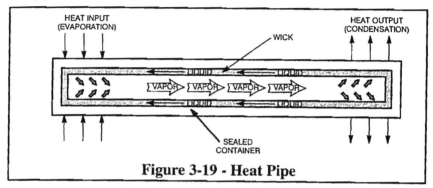

### Figure 3-19 - Heat Pipe

### 3.8.3    Thermoelectric Cooler

A thermoelectric cooler (see Figure 3-20) is a solid-state refrigerator which uses electrical current to transfer heat from a source to a hotter heat sink. Current technology thermoelectric coolers are relatively inefficient so it is important to provide a good heat sink to dissipate the power from the item being cooled plus the additional power from the thermoelectric device. In some cases the additional temperature rise of the heat sink due to the thermoelectric power negates the temperature reduction obtained from the thermoelectric device. Performance of a thermoelectric device may be analyzed [15] as follows:

$$q_c = aT_c i - \frac{\Delta T}{\theta_{te}} - \frac{1}{2}i^2 R \qquad (3\text{-}56)$$

$$I_{max} = \frac{aT_c}{R} \qquad (3\text{-}57)$$

$$Q_{c_{max}} = \frac{1}{2}a^2\frac{T_c^2}{R} \qquad (3\text{-}58)$$

$$\Delta T_{max} = \frac{1}{2}a^2\frac{\theta_{te}\,T_c^2}{R} \qquad (3\text{-}59)$$

$$\beta_{te} = \frac{2aT_c i\theta_{te} - 2\Delta T - i^2 R\theta_{te}}{2\theta_{te}(i^2 R + ai\Delta T)} \qquad (3\text{-}60)$$

$q_c$ = *Heat Transfer from Cold Side*
$\alpha$ = *Seebeck Coefficient (see text)*
$T_c$ = *Absolute Temperature of Cold Side*
$i$ = *Electrical Current Applied*
$\Delta T$ = *Temperature Difference between Hot and Cold Side*
$\theta_{te}$ = *Thermal Resistance of Thermoelectric Device (see text)*
$R$ = *Electrical Resistance of Thermoelectric Device*
$I_{max}$ = *Electrical Current for Maximum Heat Transfer*
$Q_{c_{max}}$ = *Maximum Heat Transfer (zero temperature difference)*
$\Delta T_{max}$ = *Maximum Temperature Difference (zero heat transfer)*
$\beta_{te}$ = *Coefficient of Performance of Thermoelectric (cooling rate/power)*

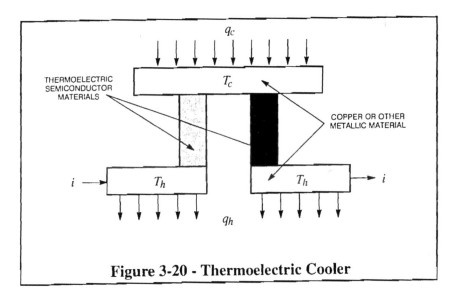

**Figure 3-20 - Thermoelectric Cooler**

The Seebeck coefficient ($\alpha$) in the above equations represents the voltage developed due to a temperature difference between the hot and cold with no electrical current flow. The thermal resistance ($\theta_{te}$) represents the thermal resistance of the thermoelectric device with no electrical current flow. It is important to note that if the thermal resistance of the thermoelectric device goes to infinity and the electrical resistance goes to zero, the thermoelectric coefficient of performance ($\beta_{te}$) matches the Carnot efficiency.

### 3.8.4    Immersion Cooling

In some cases where large amounts of power are dissipated, the components are immersed in an inert liquid to improve thermal performance. If no phase change of the fluid occur, the relationships for free convection may be applied if the appropriate material properties are used and the Grashof-Prandtl product number is within the ranges described in Table 3-2.

If phase changes (boiling) in the fluid occur, the free-convection relationships are not applicable. Information on boiling heat transfer may be found in a heat transfer textbook [12].

### 3.8.5    Phase Change Material

In some cases operation for a limited period of time without coolant is required. If the natural thermal mass of the system is not sufficient to control the temperatures over the required time interval, a phase change material may be introduced to hold a constant temperature while the material melts. The increase in operating time due to the use of a phase change material is given as follows:

$$t_a = \frac{m\ u_f}{P}$$
(3-61)

$t_a$ = Increase in Operating Time

$m$ = Mass of Phase Change Material

$u_f$ = Latent Heat of Fusion of Phase Change Material

$P$ = Net Power Delivered to Phase Change Material

## 3.9    FACTORS IN RECENT DEVELOPMENTS

Although the principles described in this chapter are applicable to a wide range of electronic packaging configurations, special consideration may need to be given to the thermal analysis of recent electronic packaging configurations (see 1.1.3). Since the packaging developments described here are at the component level, no changes in thermal analysis techniques are required at the module or chassis level. In all these situations, the guidelines and assumptions described for the appropriate conventional electronic packaging analyses apply.

### Ball-Grid Arrays (see Figure 1-5)

If the thermal conductivity of the solder balls is significantly less than that of the package, the combined effective area and thickness of the solder balls may be used in a classical analysis to determine the thermal resistance between the package and the board. This method is similar to that used to determine the interconnection thermal resistance for leadless components.

When the thermal conductivity of the solder balls is comparable to or greater than that of the package, a finite-element model (see 3.4.2) may be required to properly determine the internal component temperatures. Any finite-element model should include the contribution of any underfill.

## Multichip Modules (see Figure 1-6)

When the spacing of the chips is such that the spreading angles (see Figure 3-4) do not interfere with each other, each chip may be analyzed individually using a classical thermal analysis. Because the power dissipation in each chip may be different, it is important to apply the appropriate power level to each chip. If the thermal resistance between the package and board is relatively high, the total power dissipation of the multichip module should be used to determine the package temperature. When the thermal resistance between the package and board is relatively low, the spreading-angle approach should be continued through to the heat sink, or a finite-element approach should be used.

If the spreading angles interfere with each other, it is expected that the chips will influence each other thermally and a finite-element approach (see 3.4.2) probably will be required. If a finite element approach is used, all of the items in the overall packaging configuration expected to have an effect (leads, PWB, heat sink, etc.) should be included.

## Flip-Chips (see Figure 1-7)

Because a flip-chip configuration does not have the direct thermal path through the die bond obtained with a conventional chip attach, a finite-element model (see 3.4.2) is typically required. One possible exception to this may be the situation where a flip-chip configuration includes a heat sink applied to the back of the die (see Figure 3-21), which may be analyzed by the classical heat-spreading method. As before, all appropriate packaging parts (beam leads, solder bumps, substrate, etc.) should be included.

## Chip-on-Board (see Figure 1-8)

Chip-on-board configurations typically require finite-element (see 3.4.2) methods since the PWB has a relatively low thermal conductivity and the encapsulant may have a significant thermal effect on performance. Once again, all appropriate packaging parts (die bond, beam leads, solder bumps, PWB vias, etc.) should be included.

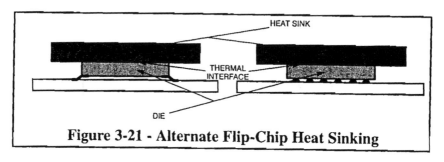

**Figure 3-21 - Alternate Flip-Chip Heat Sinking**

## 3.10   VERIFICATION

Any results from thermal analysis should be verified to help avoid any inaccuracies that might arise. This verification may be by testing, use of a simplified model, altered mesh density, or other methods as appropriate. For information on verification, see Chapter 8.

## 3.11   THERMAL ANALYSIS CHECKLIST

### 3.11.1   Applicable to All Thermal Analyses

☐   Have all appropriate thermal paths been identified?

☐   Has component power been properly defined?

☐   Have the surrounding temperatures been properly defined?

☐   Have the sources for material properties been stated?

☐   Have free convection, radiation, and forced convection been considered?

☐   Has the effect of temperature variation on material properties been considered?

☐   Has the effect of altitude been considered in free-convection situations?

☐   Has all source material (material properties, environmental data, etc.) been verified against the references?

### 3.11.2   Applicable to Transient Thermal Analysis

☐   Have all appropriate thermal masses been identified?

☐   Can each thermal mass be considered to be at a uniform temperature?

### 3.11.3  Applicable to Finite-Element Analysis

☐ Is the mesh density consistent with the heat concentrations?

☐ Has a check been made for nodes at the same location but connected to different elements?

☐ Are material types properly assigned?

### 3.11.4  Applicable to Special Heat Transfer Techniques

☐ Has the estimated power dissipation been verified?

### 3.11.5  Applicable to Recent Developments

☐ Have all relevant parts of the packaging configuration been included in the model?

### 3.11.6  Environmental Data Required

The following environmental data is typically required for various analyses. Typical units are provided for convenience, care must be taken to ensure that the system of units is consistent with the material properties and model dimensions.

**Steady-State Thermal Analysis**

☐ Boundary temperatures (°C, °F, etc.)

**Transient Thermal Analysis**

☐ Boundary temperatures (°C, °F, etc.)

☐ Duration of boundary temperatures (sec, min, hr, etc.)

**Forced-Convection Analysis**

☐ Environmental information from one of the above categories

One of the following (coolant velocity may be more appropriate for external flow situations):

☐ Coolant flow rate (kg/sec, lbm/hr, etc.)

☐ Coolant velocity (m/sec, ft/sec, etc.)

### 3.11.7  Material Properties Required

The following material properties are typically required for various analyses. Typical units are provided for convenience, care must be taken to ensure that the system of units is consistent for any analysis. If materials are not isotopic, components of various properties in the each direction may be required.

### Steady-State Thermal Analysis

☐   Thermal conductivity (W/(m-°C), Btu/(hr-ft-°F), W/(in-°C), etc.)

### Transient Thermal Analysis

☐   Material properties from above

☐   Density ($kg/m^3$, $lbm/ft^3$, etc.)

☐   Specific heat (Joule/(kg-°C), Btu/(lbm-°C), cal/(g-°C), etc.)

### Forced-Convection Analysis

☐   Properties from above (for cooling fluid)

☐   Absolute viscosity of cooling fluid (kg/(sec-m), lbm/(sec-in), etc.)

### Free-Convection Analysis

☐   Properties from above (for cooling fluid)

☐   Volume coefficient of expansion of cooling fluid (1/°C, 1/°F, etc.)

### Radiation Analysis

☐   Surface emissivity (dimensionless)

## 3.12   REFERENCES

[10]   Kays, W.M.; Convective Heat and Mass Transfer; McGraw-Hill, 1972

[11]   Lasance, C.J.M.; "Thermal Resistance: an oxymoron?", Electronics Cooling, May 1997

[12]   Holman, J.P.; Heat Transfer, Third Edition; McGraw-Hill, 1972

[13]   General Electric; Heat Transfer Data Book; General Electric Corporate Research and Development, Schenectady, NY, 1984

[14]   Chi, S.W.; Heat Pipe Theory and Practice; Hemisphere Publishing Corporation, Washington, 1976

[15]   Doolittle, J.S., Hale, F.J.; Thermodynamics for Engineers; John Wiley & Sons, 1983

[16] Fluid data files; Lockheed Martin Control Systems, Johnson City, NY; February 24, 1994

[17] Materials Engineering, 1992 Materials Selector; Penton Publishing, Cleveland, OH; December 1991

[18] General Electric; Fluid Flow Data Book; General Electric Corporate Research and Development, Schenectady, NY, 1982

[19] Dance, Francis, J; "Mounting Leadless Ceramic Chip Carriers Directly to Printed Wiring Boards - A Technology Review and Update", Circuits Manufacturing, May 1983

# Chapter 4
# Mechanical Performance
# Analysis

## 4.1  BACKGROUND

Increasing performance of aircraft, land vehicles, and other areas in which electronics are installed have increased the vibration, acoustic, shock and other mechanical loads on these systems. Any electronic system must have sufficient durability to withstand the various mechanical loads expected to be encountered over its intended life. Design of such systems requires the ability to analyze the stresses, strains, deflections, and other measurements of mechanical performance for both static and dynamic loads. These mechanical performance results are compared to material parameters to evaluate the mechanical integrity (see 4.8), or used in life analysis (see Chapter 5).

### 4.1.1  Static Mechanical Analysis

A static mechanical condition occurs when the equipment under consideration is exposed to loads that do not change over time. If the loads are steady-state, acceleration is constant. For mechanical systems static conditions occur when the duration of the mechanical loads are much larger than the period of the lowest resonant frequency. This type of condition would typically arise if a piece of equipment was required to operate under a constant acceleration for several seconds.

### 4.1.2  Dynamic Mechanical Analysis

A dynamic mechanical condition occurs when the equipment under consideration is exposed to time-varying loads. These time-varying loads can result from vibration, shock, or acoustic inputs. Dynamic conditions occur for sinusoidal vibration when the excitation frequency approaches the lowest resonant frequency of the system. For random vibration, shock, and acoustic loads, dynamic conditions occur when the highest frequency of the excitation envelope approaches the lowest system resonant frequency.

## 4.2    SELECTION OF ANALYSIS APPROACH

Thermal performance analysis uses classical analysis techniques for trade-off studies, result verification, etc. However, mechanical performance analysis (especially dynamic) requires increased complexity (usually 3 degrees of freedom instead of 1) which makes the use of classical analysis techniques difficult. Fortunately, many good finite-element packages are available which can somewhat easily provide results for static and dynamic mechanical analyses. In some cases, generic parametric finite-element models with a relatively coarse mesh can be used for initial investigations, trade-off studies, etc. These results can then be verified with more complex models using actual geometry as additional accuracy is required.

## 4.3    MECHANICAL MODELING

### 4.3.1    Background

**General Configuration**

Mechanical modeling using finite-element methods requires the modeling of the stiffness and mass that is expected to contribute to the stress, and deflection anticipated due to the applied loads. A general rule of thumb is to model to the next level back from the component under consideration. A component model would include part of the module, a chassis model would include part of the mounting system, and so on.

**Substructures/Superelements**

When modeling at higher levels of packaging, substructures and super elements may be used to account for the contribution of components and modules at the next higher level. Usually, a substructure/super element is developed by creating a model at a lower level calculating stiffness and mass matrices for the nodes that interface to the next higher level (see Figure 4-1). Typically, the overall calculation times are reduced since the number of elements that are simultaneously processed is less. For details of this technique, see Reference [23] or the user's manual of a particular finite-element software.

**Element Density**

It is important that the mesh density allows the stress pattern to be properly defined. A solid element with constant strain would not be appropriate for bending if only one element is used throughout the thickness of the part since strain is ex-

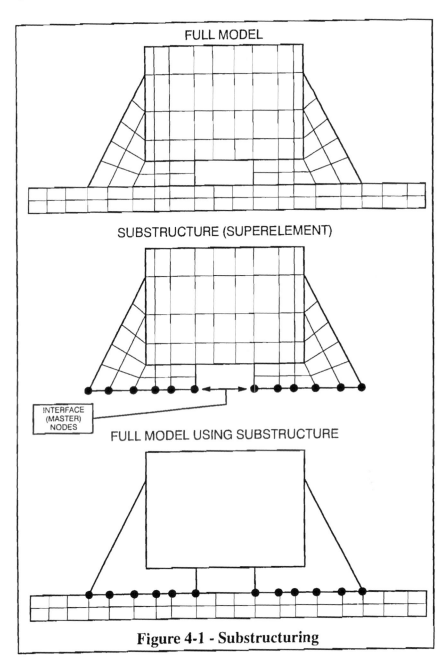

**Figure 4-1 - Substructuring**

pected to vary linearly through the thickness in bending situations. If an increased mesh density is not practical, the use of higher order elements with additional degrees of freedom or mid-side nodes (see Figure 4-2) may be used to improve accuracy. When higher order elements are used, it is important to avoid connecting the mid-side nodes of one element to the corners of another (see Reference [23] or other finite-element user's manuals).

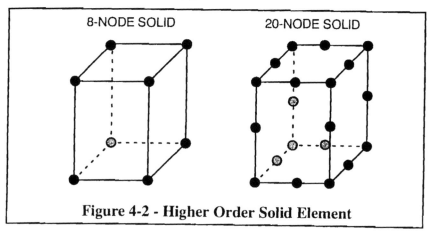

**8-NODE SOLID**          **20-NODE SOLID**

### Figure 4-2 - Higher Order Solid Element

### Element Selection

It is also important that the selection of elements be consistent with the expected stress pattern. A plate element which is designed to model bending may not be appropriate when a shear load is applied to the surface if the surface shear is not supported by the element. In any finite-element meshing application, it is important to understand the underlying stress situation to properly model the part.

### Coupling of Elements

It is also important to properly couple different types of elements. A beam element (with rotational and translational degrees of freedom) may not be attached to a solid element (with only translational degrees of freedom) without properly accounting for the relationship between the rotational degrees of freedom on the beam element and the translational degrees of freedom (DOFs) on the solid element.

### Meshing Guidelines

Typically, a finer mesh is required in the area of expected stress concentrations (corners, radii, etc.) or where concentrated loads are applied. If accurate results are not required in the area of a stress concentration, a coarser mesh may be adequate. Alternatively, the cut boundary method may be used to determine the stresses near a stress concentration. In the cut boundary method, a relatively coarse mesh is used to determine the overall deflections across a part. These deflections are then used as boundary conditions on a finer mesh model which models the area of the stress concentration (see Figure 4-3). Validity of the cut boundary approach may be verified by comparing the stresses along the boundaries for the coarse and fine models.

**Other Inconsistencies**

Occasionally a model may have inconsistencies (usually due to the modeling method) which may not be apparent with geometry plotting. Some situations to watch for include:

- Nodes at the same location but connected to different elements
- Boundary conditions that do not match the actual situation
- Boundary conditions between different levels of the analysis that are not consistent
- Material types improperly assigned
- Total mass of model does not match total system mass
- Frequency range of the analysis does not match the input environment

## 4.3.2    Component Level

Mechanical modeling of components usually requires modeling of the component body, leads, and a portion of the printed wiring board (see Figure 4-4). For simplicity and when appropriate, the leads may be modeled using beam elements, and the component body may be modeled using solid elements. The printed wiring board may be modeled using solid or plate elements, although plate elements may not be appropriate when shear loading is expected on the surface of the PWB.

## 4.3.3    Module Level

**General Configuration**

Mechanical modeling of modules usually includes the printed wiring boards (PWBs), heat sinks, and any stiffeners that are attached to the module (see Figure 4-5). The components are usually included by "smearing" the component mass over the PWB. If some components are significantly larger than the smeared mass assumption or significantly influence stiffness, they should be included separately, either as a part of the module model or a superelement generated at the component level (see 4.3.2).

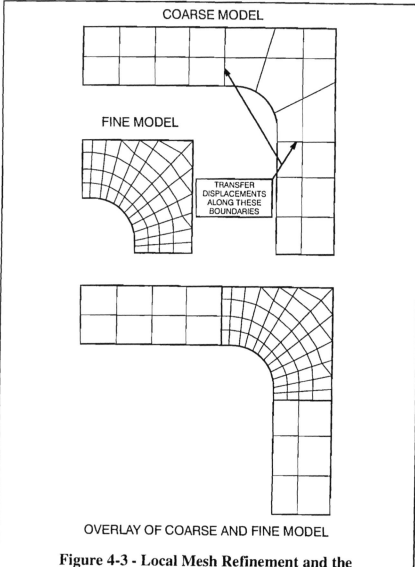

COARSE MODEL

FINE MODEL

TRANSFER
DISPLACEMENTS
ALONG THESE
BOUNDARIES

OVERLAY OF COARSE AND FINE MODEL

**Figure 4-3 - Local Mesh Refinement and the
Cut Boundary Method**

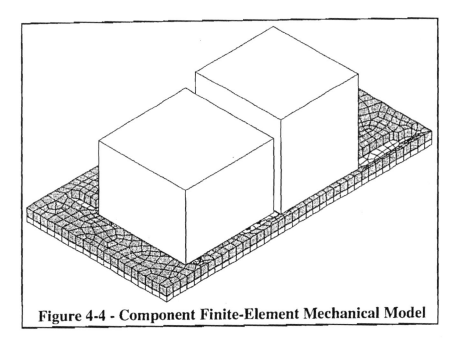

**Figure 4-4 - Component Finite-Element Mechanical Model**

### Influence of Chassis on Module

In many cases the complexity of the chassis makes the "next level back" approach difficult, so many module models use either fixed or simply supported boundary conditions. Selection of the appropriate boundary conditions depend upon the chassis stiffness and the type of module clamping method. Simply supported typically provides a "worst-case" condition which can provide some margin for error in light of various modeling assumptions. If a more detailed treatment of boundary conditions is desired, the module model may be used as a superelement in the chassis model or a chassis superelement may be used to develop module boundary conditions.

### Element Selection

It is important to reiterate that caution should be used in the selection of elements for module-level modeling, particularly if a laminated construction is used. Plate elements usually do not provide capability for shear loads on the surface and would not properly model the performance of a laminated assembly.

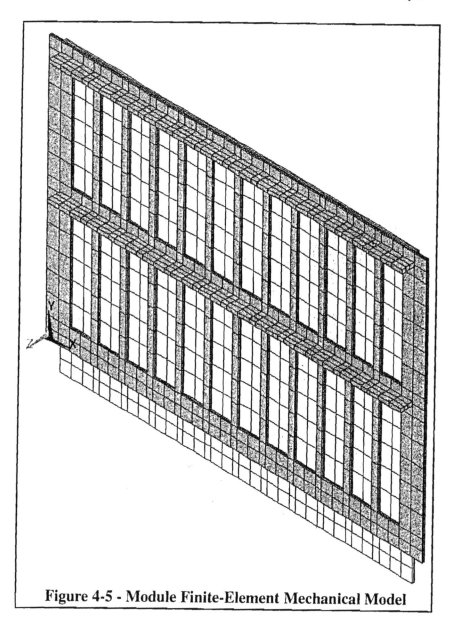

**Figure 4-5 - Module Finite-Element Mechanical Model**

### 4.3.4    Chassis Level

Modeling at the chassis level (see Figure 4-6) typically requires modeling of the basic structure and the mounting configuration. In many cases the modeling may be generated in CAD software, or an IGES file may be generated for transfer to the finite-element software. In configurations with large areas of flat sheets of material, plate elements may be appropriate although it is important to properly couple plate elements to solid elements if required. Any modules present in the chassis should be included in the model, either as approximate plates with similar stiffness and mass or as superelements.

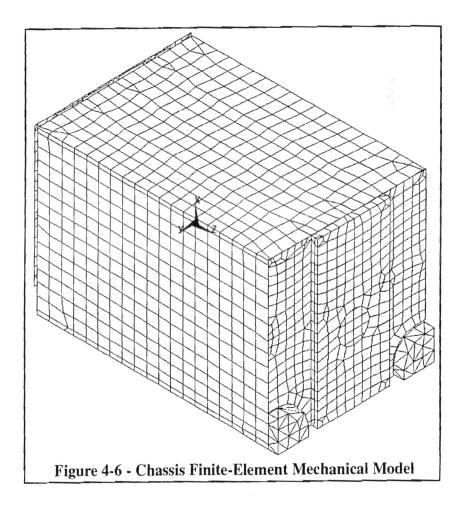

**Figure 4-6 - Chassis Finite-Element Mechanical Model**

### 4.3.5    Underlying Assumptions

- Component mass may be "smeared" over the surface of the PWB

- Contribution of the chassis at the module level may be represented by fixed or simply supported boundary conditions

# 4.4    MECHANICAL DYNAMICS BACKGROUND

### 4.4.1    Undamped Spring-Mass System

Vibration of a mechanical system is in its simplest form based upon the response of an undamped single DOF spring-mass system (see Figure 4-7). If no excitation is applied to this system, the response depends upon the initial conditions and is calculated as follows:

$$M\ddot{x} + Kx = 0 \tag{4-1}$$

$$x = D \sin(\omega t + \phi) \tag{4-2}$$

$$\dot{x} = \omega x = \omega D \cos(\omega t + \phi) \tag{4-3}$$

$$\ddot{x} = -\omega^2 x = -\omega^2 D \cos(\omega t + \phi) \tag{4-4}$$

$$\omega = 2\pi f = \sqrt{\frac{K}{M}} \tag{4-5}$$

$$x_0 = D \sin(\phi) \tag{4-6}$$

$$\dot{x}_0 = \omega D \cos(\phi) \tag{4-7}$$

$$\phi = \arctan\left(\frac{\omega\, x_0}{\dot{x}_0}\right) \quad [for\ \dot{x}_0 \geq 0]$$

$$\phi = \arctan\left(\frac{\omega\, x_0}{\dot{x}_0}\right) \pm \pi \quad [for\ \dot{x}_0 < 0] \tag{4-8}$$

$$D = \frac{x_0}{\sin(\phi)} \quad [for\ \phi \neq 0]$$

$$D = \frac{\dot{x}_0}{\omega} \quad [for\ \phi = 0] \tag{4-9}$$

$M = Mass$

$K = Spring\ Rate$

$\ddot{x} = Acceleration$

$\dot{x} = Velocity$

$x = Displacement$

$D = Response\ Amplitude$

$\omega = Circular\ Frequency$

$t = Time$

$\phi = Phase\ Angle$

$f = Vibration\ Frequency$

$x_0 = Initial\ Displacement$

$\dot{x}_0 = Initial\ Velocity$

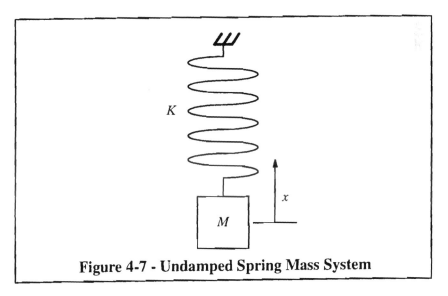

**Figure 4-7 - Undamped Spring Mass System**

## 4.4.2    Damped Spring-Mass System with Base Excitation

Typically vibration of an electronic packaging configuration consists of vibration applied to the mounting points of the hardware. If damping is applied to the single DOF spring-mass system described in 4.4.1, the configuration shown in Figure 4-8 is obtained. If steady sinusoidal base excitation is applied, the following relationships apply:

$$M\ddot{y} + C(\dot{y} - \dot{x}) + K(y - x) = 0 \tag{4-10}$$

$$x = X \sin(\omega t) \tag{4-11}$$

$$y = Y \sin(\omega t + \phi) \tag{4-12}$$

$$\omega = 2\pi f = \sqrt{\frac{K}{M}} \tag{4-13}$$

$$Q = \frac{\sqrt{K M}}{C} \tag{4-14}$$

$$A(f) = \left| \frac{Y}{X} \right| = \sqrt{\frac{1 + \left(\frac{f}{Q f_n}\right)^2}{\left(1 - \left(\frac{f}{f_n}\right)^2\right)^2 + \left(\frac{f}{Q f_n}\right)^2}} \tag{4-15}$$

$$\phi = \arctan\left(\frac{f}{Q f_n}\right) - \arctan\left[\frac{\frac{f}{Q f_n}}{1 - \left(\frac{f}{f_n}\right)^2}\right] \tag{4-16}$$

$M = Mass$

$K = Spring\ Rate$

$C = Damping$

$x, \dot{x} = Input\ Displacement,\ Velocity$

$y, \dot{y}, \ddot{y} = Response\ Displacement,\ Velocity,\ Acceleration$

$X = Input\ Amplitude$

$Y = Response\ Amplitude$

$\omega = Circular\ Frequency$

$t = Time$

$\phi = Phase\ Difference$

$f = Vibration\ Frequency$

$Q = Amplification\ at\ Resonance\ (force\ excitation,\ see\ text)$

$A(f) = Amplitude\ Transfer\ Function$

It is important to note that Q in the above equations is the amplification for force excitation which is the amplitude at resonance for a constant force excitation divided by the static deflection at the same force. Plots of amplitude transfer function vs. frequency ratio for various values of Q are shown in Figure 4-9.

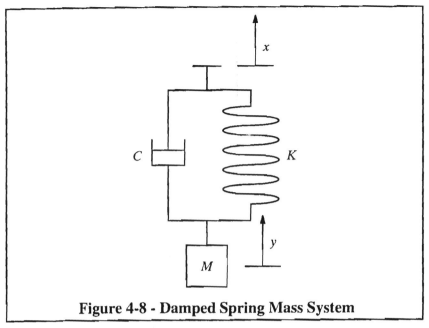

**Figure 4-8 - Damped Spring Mass System**

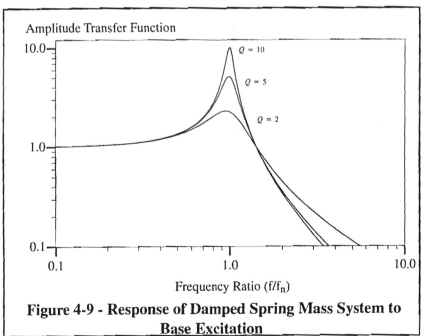

Amplitude Transfer Function

Frequency Ratio (f/f$_n$)

**Figure 4-9 - Response of Damped Spring Mass System to Base Excitation**

### 4.4.3   Random Vibration

**Background**

A random vibration environment may be simplified by considering a large number of simultaneous sinusoidal vibration inputs acting together with each input having amplitude $A_i$ and are spaced by $\Delta f$, the following relationships may be written:

$$g(t) = \sum_{i=1}^{n} [A_i \sin(\omega_i \, t + \phi_i)] \tag{4-17}$$

$$n = \frac{(f_h - f_l)}{\Delta f} \tag{4-18}$$

$$\omega_i = 2\pi \left( f_l + i \, \Delta f - \frac{\Delta f}{2} \right) \tag{4-19}$$

$$g_{RMS}^2 = \frac{1}{T} \int_0^T \left( \sum_{i=1}^{n} [A_i \sin(\omega_i \, t + \phi_i)]^2 \right) dt \tag{4-20}$$

$$g_{RMS}^2 = \sum_{i=1}^{n} \left[ A_i^2 \lim_{T \to \infty} \left( \frac{1}{T} \int_0^T \sin^2(\omega_i \, t + \phi_i) dt \right) \right] = \sum_{i=1}^{n} \left( A_i^2 \frac{1}{2} \right) \tag{4-21}$$

$$g_{RMS}^2 = n \left( \frac{A_i^2}{2} \right) = \frac{(f_h - f_l)}{\Delta f} \left( \frac{A_i^2}{2} \right) = (f_h - f_l) \, PSD(f) \tag{4-22}$$

Solving for A:

$$A_i = \sqrt{2 \, PSD(f) \, \Delta f} \tag{4-23}$$

$g(t) = $ *Random Signal (time domain)*

$t = $ *Time*

$n = $ *Number of Inputs*

$i = $ *Input Counter*

$A_i = $ *Amplitude of Each Input*

$\omega_i = $ *Circular Frequency*

$\phi_i = $ *Random Phase*

$f_h = $ *High Limit of Random Spectrum*

$f_l = $ *Low Limit of Random Spectrum*

$\Delta f$ = *Spacing between Inputs*

$g_{RMS}$ = *Root Mean Square Acceleration*

$T$ = *Integration Time*

$PSD(f)$ = *Power Spectral Density (function of frequency)*

## Distribution of Amplitudes

Amplitude of the time-domain random environment is normally distributed with a standard deviation ($\sigma$) equal to the RMS value leading to the distribution of amplitudes as shown in Table 4-1. This distribution of amplitudes over time becomes significant in life analyses (see Chapter 5).

If a uniform random signal is applied to a single DOF spring-mass system (see Figure 4-8) the RMS response is calculated as follows:

$$A(f) = \sqrt{\frac{1 + \left(\frac{f}{Qf_n}\right)^2}{\left(1 - \left(\frac{f}{f_n}\right)^2\right)^2 + \left(\frac{f}{Qf_n}\right)^2}} \tag{4-15}$$

$$g_{RMS}^2 = \int_0^\infty PSD \ (A(f))^2 \ df \tag{4-24}$$

$$g_{RMS}^2 = PSD \int_0^\infty \frac{1 + \left(\frac{f}{Qf_n}\right)^2}{\left(1 - \left(\frac{f}{f_n}\right)^2\right)^2 + \left(\frac{f}{Qf_n}\right)^2} \ df = \frac{\pi}{2} \frac{1 + Q^2}{Q} f_n \ PSD \tag{4-25}$$

$$g_{RMS} = \sqrt{\frac{\pi}{2} \frac{1 + Q^2}{Q} f_n \ PSD} \tag{4-26}$$

$$\delta_{RMS} = \frac{g_{RMS}}{(2\pi f_n)^2} = \frac{\sqrt{\frac{\pi}{2} \frac{1 + Q^2}{Q} f_n \ PSD}}{(2\pi f_n)^2} \tag{4-27}$$

$A(f)$ = *Amplitude Transfer Function*

$f$ = *Frequency*

$f_n$ = *Resonant Frequency*

$Q$ = *Amplfication at Resonance (force excitation, see text)*

$g_{RMS}$ = *RMS Acceleration*

$PSD$ = *Power Spectral Density (uniform)*

$\delta_{RMS}$ = *RMS Deflection*

Care should be taken to ensure that the proper conversion for deflection and acceleration are used when the power spectral density is expressed in terms of $g^2$/Hz.

| Table 4-1 - Amplitude Distribution of Random Vibration ||
| Amplitude Range | Fraction of Time |
|:---:|:---:|
| < 1 x RMS | 68.26% |
| 1 x RMS – 2 x RMS | 27.18% |
| 2 x RMS – 3 x RMS | 4.30% |
| > 3 x RMS | 0.26% |

**Relative vs. Absolute Displacement**

It is important to note that the RMS acceleration given in equation 4-26 and displacement given in equation 4-27 are absolute values relative to the static (not vibrated) reference system. Displacements that produce stress in various components need to be evaluated relative to the input to that component. Many finite-element packages have the capability to evaluate relative displacements. Proper accounting of relative displacements must also be considered in the analysis of vibration test data (for details see 7.3.3). For the single DOF system described above, the relative displacement is calculated as follows:

$$R(f) = \left| \frac{Y-X}{X} \right| = \frac{\left(\frac{f}{f_n}\right)^2}{\sqrt{\left(1 - \left(\frac{f}{f_n}\right)^2\right)^2 + \left(\frac{f}{Qf_n}\right)^2}} \tag{4-28}$$

$$g'_{RMS}{}^2 = \int_0^\infty PSD \, (R(f))^2 \, df \tag{4-29}$$

$$\delta'_{RMS}{}^2 = \int_0^\infty PSD \left(\frac{R(f)}{(2\pi f)^2}\right)^2 df \qquad (4\text{-}30)$$

$$\delta'_{RMS}{}^2 = \int_0^\infty PSD \left[\frac{\left(\frac{f}{Qf_n}\right)^4}{\left(1 - \left(\frac{f}{f_n}\right)^2\right)^2 + \left(\frac{f}{Qf_n}\right)^2}\right] \frac{1}{(2\pi f)^4} df \qquad (4\text{-}31)$$

$$\delta'_{RMS}{}^2 = \frac{Q \cdot PSD}{32 \ \pi^3 f_n{}^3}$$

$$\delta'_{RMS} = \sqrt{\frac{Q \cdot PSD}{32 \ \pi^3 f_n{}^3}} \qquad (4\text{-}32)$$

$R(f)$ = *Relative Amplitude Transfer Function*
$X$ = *Input Amplitude*
$Y$ = *Response Amplitude*
$f$ = *Frequency*
$f_n$ = *Resonant Frequency*
$Q$ = *Amplification at Resonance (force excitation, see text)*
$g'_{RMS}$ = *Relative RMS Acceleration*
$PSD$ = *Power Spectral Density*
$\delta'_{RMS}$ = *RMS Deflection*

As was noted previously, care should be taken to ensure that the proper conversion for deflection and acceleration are used when the power spectral density is expressed in terms of $g^2$/Hz.

## 4.4.4 Acoustic Noise

**Description of Acoustic Environments**

The effect of acoustic energy on a surface may be expressed as a time-varying pressure that develops deflection and stress on a surface [25,26]. Typically the magnitude of acoustic energy is expressed as a sound pressure level (SPL) and measured in decibels relative to a reference pressure as follows:

$$SPL = 20 \log\left(\frac{p}{p_0}\right) \qquad (4\text{-}33)$$

$SPL$ = *Sound Pressure Level (db)*

$p$ = *Pressure Amplitude*

$p_0$ = *Reference Pressure* = 0.0002 *μbar (in air)*

In the analysis of the effects of acoustic energy on hardware, it is important to consider the frequency distribution of this acoustic energy. The magnitude of the SPL for a broad-band acoustic signal increases with increasing bandwidth, so measurements of broad-band acoustic pressures must include consideration for frequency. This acoustic pressure is usually expressed as a pressure spectrum level (PSL) or a pressure band level (PBL) over a frequency range (see Figure 4-10). A PSL is the SPL contained within a frequency band 1 Hz wide and a PBL is the SPL in a frequency band Δf wide. The relationship between the PBL and PSL is given as follows:

$$PBL = PSL + 10 \log(\varDelta f) \qquad (4\text{-}34)$$

$PBL$ = *Pressure Band Level (db)*

$PSL$ = *Pressure Spectrum Level (db)*

$\varDelta f$ = *Band Width*

The band width for PBL spectrum definitions is typically expressed relative to an octave (doubling of frequency) such as one-third octave or half octave. Figure 4-10 shows the same acoustic environment expressed as pressure spectrum level and pressure band level for various band widths.

### Calculation of Pressure Spectral Density

Pressure spectral density (analogous to power spectral density) provides a convenient method for calculating the input environment in a method that is independent of the selection of pressure band widths. Pressure band level (PBL) values may be converted to pressure spectral density values as follows:

For a linear variation of PBL:

$$PBL = A + B \log(f) \qquad (4\text{-}35)$$

$$B = \frac{M}{\log(2)} \qquad (4\text{-}36)$$

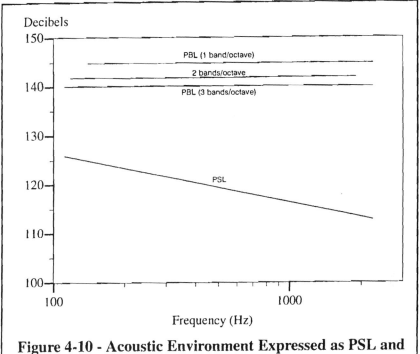

**Figure 4-10 - Acoustic Environment Expressed as PSL and PBL with Various Band Widths**

$$A = PBL_1 - M \frac{\log(f_1)}{\log(2)} \tag{4-37}$$

$$p = p_0 \ 10^{\left(\frac{PBL}{20}\right)} \tag{4-38}$$

$$p = p_0 \ 10^{\left(\frac{A + B \log(f)}{20}\right)} \tag{4-39}$$

$$PrSD = \frac{p^2}{\Delta f} \tag{4-40}$$

$$PrSD = \frac{\left[p_0 \ 10^{\left(\frac{A + B \log(f)}{20}\right)}\right]^2}{K f} \tag{4-41}$$

$$\Delta f = K f \tag{4-42}$$

$$PrSD = \frac{p_0^2 \; 10^{\left(\frac{A}{10}\right)}}{K} f^{\left(\frac{B}{10}-1\right)} \tag{4-43}$$

$$PSL = 10 \log\left(\frac{PrSD}{p_0^2}\right) \tag{4-44}$$

$PBL$ = Pressure Band Level (db)
$A, B$ = Constants
$f$ = Frequency
$M$ = Slope of Band SPL (db/octave)
$PBL_1$ = PBL at Frequency $f_1$
$p$ = Band Pressure Amplitude
$p_0$ = Reference Pressure = 0.0002 $\mu bar$ (in air)
$PrSD$ = Pressure Spectal Density
$\Delta f$ = Bandwidth
$K$ = Bandwidth Factor (see text and Table 4-2)
$PSL$ = Pressure Spectrum Level

The bandwidth factor (K) represents the ratio between the bandwidth at a given frequency and that frequency. For typical one-third octave bands, the value is 0.2316 and values for other bands per octave may be found in Table 4-2.

| Table 4-2 - Bandwidth Factor for Various Bands per Octave | |
|:---:|:---:|
| **Bands per Octave** | **Band Width Factor (K)** |
| 1 | 0.7071 |
| 2 | 0.3483 |
| 3 | 0.2316 |
| 4 | 0.1735 |
| N | $2^{\left(\frac{1}{2N}\right)} - 2^{\left(-\frac{1}{2N}\right)}$ |

**Overall Sound Pressure Level**

The pressure spectral density curve can be used to determine the overall sound pressure level (SPL) as follows:

$$p = \sqrt{\left[\int_{f_l}^{f_h} PrSD(f)\ df\right]} \qquad (4\text{-}45)$$

$$SPL = 20 \log\left(\frac{p}{p_0}\right) \qquad (4\text{-}33)$$

$p$ = RMS Pressure Amplitude
$PrSD(f)$ = Pressure Spectral Density (function of frequency)
$f_l, f_h$ = Low and High Limits of Frequency Range
$SPL$ = Overall Sound Pressure Level (db)
$p_0$ = Reference Pressure = 0.0002 $\mu bar$ (in air)

## Acoustic Intensity due to Wall Motion

The time-varying pressure that acoustic energy produces results in deflection of bodies to which the acoustic energy is applied. If the body exposed to the acoustic energy is hollow, like an electronic chassis, the motion of the wall conversely produces a pressure which results in acoustic intensity within the body. If the acoustic response is desired for a location very close to the moving wall, the deflection can be used to determine the pressure amplitude as follows:

$$p = \rho_0 c\ u \qquad (4\text{-}46)$$

$$p = \rho_0 c\ 2\pi f \delta \qquad (4\text{-}47)$$

$p$ = Pressure Amplitude
$\rho_0$ = Equilibrium Density of Fluid
$c$ = Speed of Sound in Fluid
$\rho_0 c$ = Specific Acoustic Impedance
$u$ = Particle Velocity
$f$ = Frequency of Oscillation
$\delta$ = Wall Deflection

## Acoustic Transmission

Acoustic energy impinging from air upon a wall and back into air results in some energy being transmitted and some energy being reflected. If the wall has a high specific acoustic impedance relative to air (such as a metal) and is relatively thin the attenuation due to the wall is calculated as follows:

$$a_t = \frac{4(\rho_a c_a)^2}{(\rho_w c_w)^2 (k_w l)^2} \tag{4-48}$$

$$k_w = \frac{\omega}{c_w} = \frac{2\pi f}{c_w} \tag{4-49}$$

$$L = -10 \log(a_t) \tag{4-50}$$

$$p_t = p_i \sqrt{a_t} \tag{4-51}$$

$a_t$ = *Sound Power Transmission Coefficient*
$\rho_a c_a$, $\rho_w c_w$ = *Specific Acoustic Impedance in Air, Wall*
$k_w$ = *Wavelength Constant in Wall*
$l$ = *Wall Thickness*
$\omega$ = *Circular Frequency*
$f$ = *Frequency of Oscillation*
$c_w$ = *Speed of Sound in Wall*
$L$ = *Attenuation (Loss) in Decibels*
$p_t$, $p_i$ = *Transmitted, Incident Pressure Amplitude*

Conditions:

$$\rho_w c_w \gg \rho_a c_a \text{ (high relative impedance)}$$
$$k_w l \ll 1 \text{ (thin wall)}$$

The above relationship for attenuation due to transmission may be used for pressure spectral densities as follows:

$$PrSD_t = a_t \, PrSD_i \tag{4-52}$$

$PrSD_t$, $PrSD_i$ = *Transmitted, Incident Pressure Spectral Densities*

**Underlying Assumptions**

- Wave motion near a wall may be represented by acoustic plane waves
- Specific acoustic impedance of the wall is much higher than air
- Wall is relatively thin

## 4.4.5   Mechanical Shock

Analogous to vibration of electronic packaging configurations, mechanical shock consists of a shock pulse applied to the mounting points of the hardware. One method of calculating shock response is to convert the input pulse into the frequency domain using a fourier series as follows:

$$x(t) = \frac{1}{2}A_0 + \sum_{n=1}^{\infty} C_n \sin(n\omega t + \phi_n) \tag{4-53}$$

$$C_n = \sqrt{A_n^2 + B_n^2} \tag{4-54}$$

$$\phi_n = \arctan\left(\frac{A_n}{B_n}\right) \quad [for\ A_n \geq 0]$$

$$\phi_n = \arctan\left(\frac{A_n}{B_n}\right) \pm \pi \quad [for\ A_n < 0] \tag{4-55}$$

$$A_n = \frac{2}{T}\int_0^T x(t)\cos(n\omega t) \tag{4-56}$$

$$B_n = \frac{2}{T}\int_0^T x(t)\sin(n\omega t) \tag{4-57}$$

$$\omega = \frac{2\pi}{T} \tag{4-58}$$

$x(t)$ = Time Domain Signal
$C_n$ = Magnitude of nth Harmonic
$A_n, B_n$ = Cosine, Sine Terms of nth Harmonic
$n$ = Positive Integer
$\omega$ = Circular Frequency
$t$ = Time
$\phi_n$ = Phase Angle of nth Harmonic
$T$ = Period of Time Domain Signal

The amplitude and phase relationships for the response transfer function can be applied to the frequency domain representation to determine the shock response. For the single DOF system shown in Figure 4-8, equations 4-15 and 4-16 give the amplitude and phase response.

The shock response of a system may also be calculated in the time domain using a finite difference method. In this method, the equations of motion are used to deter-

mine the response on a finite time-slice basis. The conditions at the beginning of
the time-slice and the calculated derivatives are used to predict the conditions at
the end of the time-slice. For the damped spring mass system with base excitation
(Figure 4-8) the response is calculated as follows:

$$M\ddot{y} + C(\dot{y} - \dot{x}) + K(y - x) = 0 \qquad (4\text{-}10)$$

$$\ddot{y} = \frac{C}{M}(\dot{x} - \dot{y}) + \frac{K}{M}(x - y) \qquad (4\text{-}59)$$

$$\ddot{y}_n = \frac{C}{M}(\dot{x}_n - \dot{y}_n) + \frac{K}{M}(x_n - y_n) \qquad (4\text{-}60)$$

$$\dot{y}_{n+1} = \dot{y}_n + \ddot{y}_n \, \Delta t \qquad (4\text{-}61)$$

$$y_{n+1} = y_n + \frac{(\dot{y}_{n+1} + \dot{y}_n)}{2} \, \Delta t \qquad (4\text{-}62)$$

$$t_{n+1} = t_n + \Delta t \qquad (4\text{-}63)$$

$M = Mass$

$K = Spring\ Rate$

$C = Damping$

$x, \dot{x} = Input\ Displacement,\ Velocity$

$y, \dot{y}, \ddot{y} = Response\ Displacement,\ Velocity,\ Acceleration$

$x_n \dot{x}_n = Input\ Displacement,\ Velocity\ at\ Step\ n$

$y_n \dot{y}_n, \ddot{y}_n = Response\ Displacement,\ Velocity,\ Acceleration\ at\ Step\ n$

$y_{n+1}, \dot{y}_{n+1} = Response\ Displacement,\ Velocity\ at\ Step\ n+1$

$\Delta t = Time\ Increment$

$t_{n+1}, t_n = Time\ at\ Step\ n+1,\ n$

Typical response to a half-sine shock calculated using the above method is shown
in Figure 4-11. More sophisticated methods for finite difference solutions of differential equations (such as Runge-Kutta) are available, but the above method
serves to explain the concept and can produce adequate results if the time increment is properly chosen.

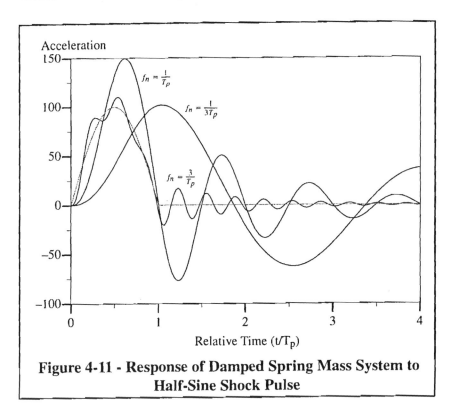

Figure 4-11 - Response of Damped Spring Mass System to
Half-Sine Shock Pulse

# 4.5    DYNAMIC MECHANICAL ANALYSIS

## 4.5.1    Modal Analysis

Usually any vibration analysis starts with a modal analysis to calculate the resonant frequencies and mode shapes of a system. In mathematical terms, the geometry is expressed as mass and stiffness matrices. The equations of motion for the
system are then expressed in a matrix form analogous to the undamped spring-
mass system described in 4.4.1:

$$[M](\ddot{u}) + [K](u) = (0) \tag{4-64}$$

$$-[M]\omega^2(u) + [K](u) = (0) \tag{4-65}$$

$$\omega = 2\pi f \tag{4-66}$$

$$\big([K] - [M]\omega^2\big)(u) = (0) \tag{4-67}$$

$[M]$ = *Mass Matrix*

$[K]$ = *Stiffness Matrix*

$(\ddot{u})$ = *Acceleration Vector*

$(u)$ = *Displacement Vector*

$\omega$ = *Circular Frequency*

$f$ = *Vibration Frequency*

The equation 4-67 is a classical eigenvalue problem [20] in which the eigenvalues represent the square of the circular frequency and the eigenvectors represent the mode shapes. Figure 4-12 illustrates a typical mode shape obtained from finite-element analysis of an electronic module.

## 4.5.2    Sinusoidal Vibration Analysis

Sinusoidal vibration analysis of a finite-element model requires calculating the harmonic response of a multiple degree of freedom system in a method analogous to the damped spring mass response calculation described in 4.4.2. Most finite-element software packages provide the capability to perform harmonic response analysis. Proper characterization of damping is important for calculating the responses of the system, particularly near resonances.

## 4.5.3    Random Vibration Analysis

### Background

Random vibration analysis of a finite-element model is a probabalistic method which determines the root mean square displacements, stresses, etc. based upon an input power spectral density. Many finite-element software packages provide the capability to calculate response to random excitation spectra. Details on the theoretical basis for calculating random spectrum response of a finite-element model may be found in Reference [22] or the theoretical manual of a particular finite-element software.

### Calculation of Curvatures in Random Vibration Analysis [21]

In some cases the curvature of a PWB or similar geometry is required for subsequent analyses (such as solder-life analysis, see 5.2.2). Some finite-element codes do not provide for output of curvature so the curvatures need to be calculated from the strains on the surfaces of the PWB. In sinusoidal vibration and static models, the curvature is the difference in strain between the top and bottom surface of the PWB divided by the thickness. For random vibration, the strains represent the RMS amplitude which is a statistical and not vector quantity. Since the RMS am-

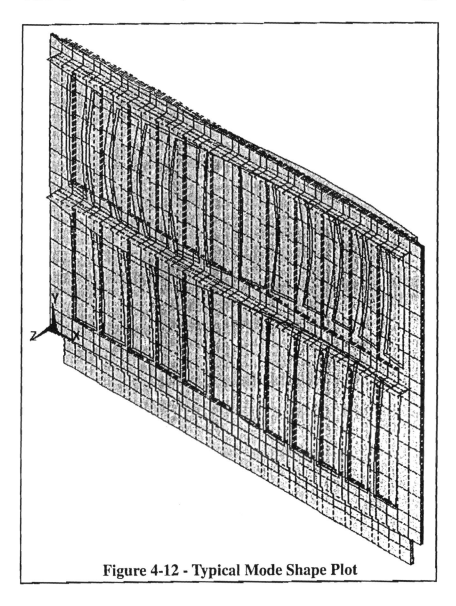

**Figure 4-12 - Typical Mode Shape Plot**

plitudes are not vector quantities, particular attention must be paid to properly calculating the RMS curvature. For a PWB under random vibration, the RMS curvature of the module surface is calculated as follows:

$$(\mathcal{C}_{x,y})_{RMS} = \sqrt{\sum_{i=1}^{n}\sum_{j=1}^{n}\left(\frac{\epsilon_{f_{x,y}} - \epsilon_{b_{x,y}}}{t}\right)_i\left(\frac{\epsilon_{f_{x,y}} - \epsilon_{b_{x,y}}}{t}\right)_j Q_{ij}} \qquad (4\text{-}68)$$

$$(\mathcal{C}_{x,y})_{RMS} = \frac{\sqrt{\sum_{i=1}^{n}\sum_{j=1}^{n}\left(\epsilon_{f_{x,y}} - \epsilon_{b_{x,y}}\right)_i\left(\epsilon_{f_{x,y}} - \epsilon_{b_{x,y}}\right)_j Q_{ij}}}{t} \qquad (4\text{-}69)$$

$\mathcal{C}_{x,y}$ = Curvature in x,y Direction

$(\mathcal{C}_{x,y})_{RMS}$ = RMS Quantity

$(\mathcal{C}_{x,y})_{i,j}$ = Mode Shape i,j Quantity

$\epsilon_{f_{x,y}}$ = PWB Surface Strain on Front Side in x,y Direction

$\epsilon_{b_{x,y}}$ = PWB Surface Strain on Back Side in x,y Direction

$Q_{ij}$ = Modal Covariance Matrix (from finite-element analysis)

$t$ = PWB Thickness

## Use of Harmonic Response Data for Random Vibration Analysis

If the capability to calculate response to a random spectrum is not available in a finite-element software package, the responses may be calculated as follows:

1. Conduct a harmonic response analysis over the frequency range of the random excitation spectrum using a frequency increment that accommodates the appropriate resonant frequencies, and amplitude as given by equation 4-23

2. RMS responses are determined by taking the square root of the sum of squares of the appropriate stresses, deflections, etc.

3. Deflections, stresses, etc. may be converted to deflection, and stress spectral density as follows:

$$DSD_i = \frac{\delta_i^2}{2\,\Delta f} \qquad (4\text{-}70)$$

$$SSD_i = \frac{\sigma_i^2}{2\,\Delta f} \qquad (4\text{-}71)$$

$\delta_i$ = *Peak Deflection at Frequency i*
$\Delta f$ = *Frequency Increment*
$DSD_i$ = *Deflection Spectral Density at Frequency i*
$\sigma_i$ = *Peak Stress at Frequency i*
$SSD_i$ = *Deflection Spectral Density at Frequency i*

## 4.5.4    Acoustic Noise Analysis

**Acoustic Response**

The approach for determining the stresses and deflections resulting from acoustic noise is as follows:

1. Perform a modal harmonic response analysis using a constant harmonically varying pressure input

2. Use the pressure spectral density in conjunction with the width of each frequency band to determine the input pressure at each frequency.

3. Post process the harmonic response stresses and/or deflections by using the pressure from step 2 to scale the results at each frequency

4. Root of sum of squares (RSS) add results from step 3 over the frequency range to determine overall root mean square (RMS) response.

The relationship between stresses and/or deflections to input pressure can be used to develop deflection and/or stress spectral density response curves for visualization of acoustic response. It is important to perform the harmonic response analysis described in step 1 over the full frequency range corresponding to the acoustic input.

**Acoustic Intensity Inside Chassis**

The acoustic intensity inside the chassis is calculated by adding the intensities of acoustic energy due to wall motion and transmission. It is important to note that both the intensity due to wall motion and that due to transmission vary with frequency so proper consideration must be made for frequency dependence. The intensity due to wall motion also varies over the surface but a worst-case estimate may be made by using the maximum wall deflection. If a more exact calculation is desired, a finite-element model may be made of the chassis wall and the air to determine the acoustic pressure distribution inside the chassis.

**Underlying Assumptions**

- Maximum wall deflection is used to determine worst-case acoustic intensity inside chassis

### 4.5.5    Mechanical Shock Analysis

Mechanical shock analysis of a finite-element model requires calculating the transient response of a multiple degree of freedom system in a method analogous to the damped spring mass response calculation described in 4.4.5. The method used by the finite-element software may be based upon a frequency domain solution using a Fourier analysis of the shock pulse in conjunction with the harmonic response of the structure, or may be a true transient calculation similar to the finite difference method.

## 4.6    STATIC MECHANICAL ANALYSIS

### 4.6.1    Lead Stiffness

Lead stiffness analysis requires applying the required boundary conditions and determining the appropriate forces and stresses. If the lead is very stiff, the effect of the solder joint should be considered. As in any analysis, it is important that the boundary conditions closely match the actual situation.

### 4.6.2    Sustained Acceleration

Analysis of sustained acceleration performance requires applying the inertia loads to the finite-element model and performing a static solution. In less complex systems where finite-element solutions are not required, such as mounting foot loads, the sustained acceleration solution may be obtained by performing a static force balance on the system. In such cases, proper consideration of the mass and location of the center of gravity is important.

### 4.6.3    Thermal Stresses/Strains

**Thermal Stress Analysis**

Thermal stresses/strains occur when differences in thermal expansion are encountered within an assembly. These differences in thermal expansion may arise when materials with varying coefficients of thermal expansion (CTEs) are exposed to a constant temperature change or a homogeneous material is exposed to a temperature gradient. For a simple case with two materials (see Figure 4-13), the thermal stresses/strains and CTEs are calculated as follows:

$$\epsilon_a = \epsilon_{t_1} + \epsilon_{m_1} = \epsilon_{t_2} + \epsilon_{m_2} \qquad (4\text{-}72)$$

$$\epsilon_{t_{1,2}} = \alpha_{1,2} \; ( \; T_f - T_i \; ) \qquad (4\text{-}73)$$

$$\epsilon_{m_{1,2}} = \frac{\sigma_{1,2}}{E_{1,2}} \qquad (4\text{-}74)$$

$$\sigma_1 A_1 + \sigma_2 A_2 = 0 \qquad (4\text{-}75)$$

$$\sigma_1 = \Delta T(\alpha_2 - \alpha_1)\frac{A_2 \; E_1 \; E_2}{A_1 \; E_1 + A_2 \; E_2} \qquad (4\text{-}76)$$

$$\sigma_2 = \Delta T(\alpha_1 - \alpha_2)\frac{A_1 \; E_1 \; E_2}{A_1 \; E_1 + A_2 \; E_2} \qquad (4\text{-}77)$$

$$\Delta T = T_f - T_i \qquad (4\text{-}78)$$

$$\alpha_a = \frac{\epsilon_a}{\Delta T} = \frac{\epsilon_{t_1} + \epsilon_{m_1}}{\Delta T} = \frac{\epsilon_{t_2} + \epsilon_{m_2}}{\Delta T} \qquad (4\text{-}79)$$

$\epsilon_a = $ *Overall Strain of Assembly*
$\epsilon_{t_{1,2}} = $ *Thermal Expansion of Material 1,2*
$\epsilon_{m_{1,2}} = $ *Mechanical Strain of Material 1,2*
$\alpha_{1,2} = $ *Coefficient of Thermal Expansion (CTE) of Material 1,2*
$T_f = $ *Final Temperature*
$T_i = $ *Initial Temperature*
$\sigma_{1,2} = $ *Stress in Material 1,2*
$E_{1,2} = $ *Modulus of Elasticity of Material 1,2*
$A_{1,2} = $ *Total Cross Sectional Area of Material 1,2*
$\Delta T = $ *Temperature Change*
$\alpha_a = $ *Overall CTE of Assembly*

In more complex cases, finite-element approaches are usually required. Most fi-nite-element codes provide the capability to input the material CTEs and the ex-pected temperature distribution to calculate the resulting stresses and strains. In some cases where the temperature is not uniform, the temperature distribution may be calculated using the finite-element method.

## Thermal Shock/Gradient Stresses

As was described previously, a non-uniform temperature distribution can induce thermal stresses. This situation can occur when a temperature gradient or a sudden temperature change is applied to an assembly. This situation can be analyzed us-

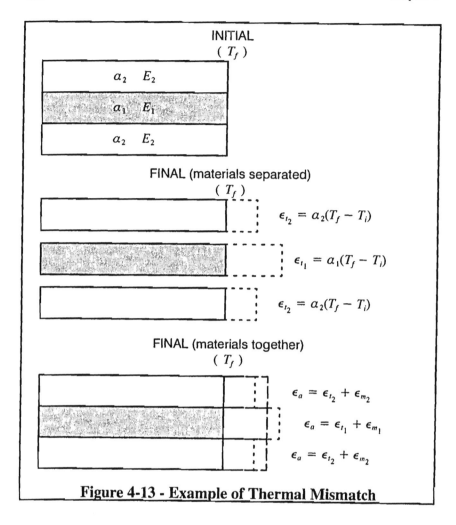

**Figure 4-13 - Example of Thermal Mismatch**

ing a finite-element code that has both thermal and mechanical (thermal stress) analysis capability. Specifically, the procedure for calculating thermal stresses in this situation is as follows:

1.  Generate a finite-element thermal model of the assembly. Include both thermal and mechanical material properties.

2.  Perform a transient thermal analysis using the appropriate thermal boundary conditions. If the expected stresses will be due to a steady-state temperature gradient, then a steady-state analysis may be adequate.

3.  Change the thermal elements to mechanical elements.

4.  Conduct a static mechanical analysis using the temperatures calculated in step 2. Since the thermal time constant is typically much greater than the period of mechanical oscillation, a dynamic analysis is not required.

A typical finite-element model used to calculate thermal stresses is shown in Figure 4-14.

## 4.7  TYPICAL ANALYSES

### 4.7.1  Typical Component Lead Stiffness Model

Life analysis often requires the stiffness and stress factors of the component leads to be determined. For the copper lead geometry shown in Figure 4-15 determine the following:

- Lead stiffness

- Lead stress factors

**Development of Finite-Element Model**

A finite-element model is generated for the lead with a typical solder joint using 3,479 nodes, and 1,440 elements for the lead and 1,200 elements for the solder (see Figure 4-16). The solder pad dimensions are 0.075 in x 0.030 in with a 0.003-in minimum solder thickness. Fixed constraints are applied at the indicated surface (see Figure 4-15), and material properties are as shown in Table 4-3.

**Stiffness Matrix Determination**

Deflection of 0.001 in is applied independently (constrained to zero in other directions) in each of the coordinate directions, resulting in the following forces:

|  | 0.001-in Deflection Direction | | |
|---|---|---|---|
| Force (lbf) | X (axial) | Y (vertical) | Z (transverse) |
| X (axial) | 20.868 | 24.821 | 0 |
| Y (vertical) | 24.821 | 57.517 | 0 |
| Z (transverse) | 0 | 0 | 13.065 |

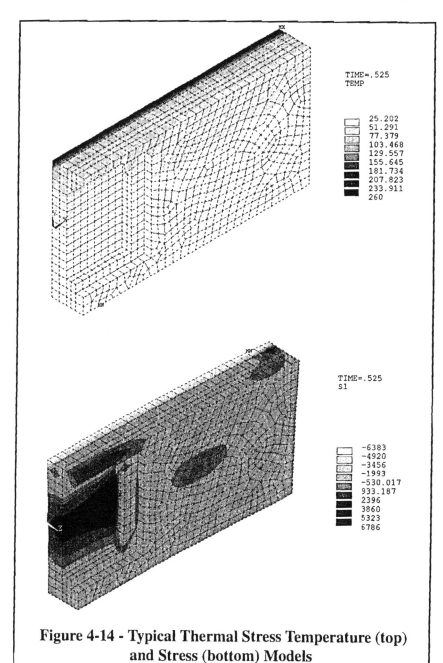

**Figure 4-14 - Typical Thermal Stress Temperature (top)
and Stress (bottom) Models**

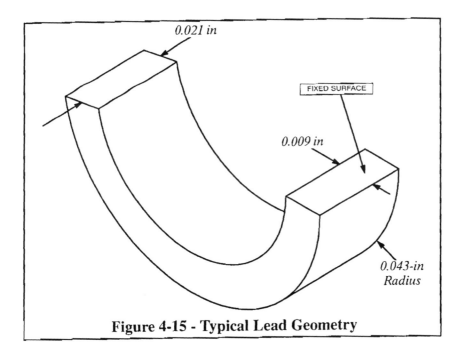

**Figure 4-15 - Typical Lead Geometry**

Dividing the above forces by the input deflection results in the following stiffness matrix components:

$$K_a = 20,868 \frac{lbf}{in} \qquad\qquad K_v = 57,517 \frac{lbf}{in}$$

$$K_{av} = K_{va} = 24,821 \frac{lbf}{in} \qquad\qquad K_t = 13,065 \frac{lbf}{in}$$

$K_{a,v,t}$ = Diagonal Terms in Axial, Vertical, and Transverse Directions

$K_{av,va}$ = Vertical to Axial Stiffness

$K_{vt,tv,at,ta}$ = 0 = Transverse to Axial and Vertical Stifness

**Lead Stress Factors**

Von Mises lead stresses due to the 1-mil deflections in axial, transverse, and vertical directions are shown in Figures 4-17 through 4-19.

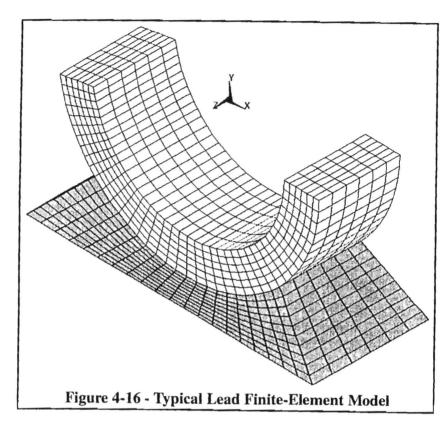

**Figure 4-16 - Typical Lead Finite-Element Model**

## Table 4-3 - Material Properties Used for Typical Analyses

| Material [source] | Modulus of Elasticity (psi) | Poisson's Ratio | Thermal Expansion (ppm/°C) | Density (lbm/in³) |
|---|---|---|---|---|
| Aluminum [27][28] | $10 \times 10^6$ | 0.33 | 23.4 | 0.098 |
| Copper [27][28] | $17.5 \times 10^6$ | 0.345 | - | - |
| Solder [29] | $3.6 \times 10^6$ | 0.35 | 28.3 | - |
| Glass/epoxy PWB [30] | $2.5 \times 10^6$ | - | 14.4 | 0.070 |
| Silicone rubber [27][28] | 500 | 0.495 | 160 | 0.040 |

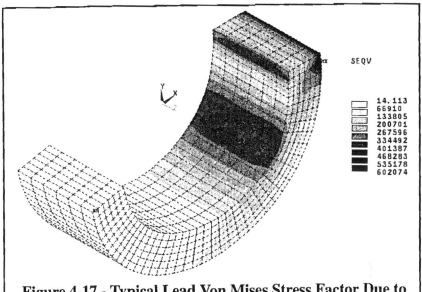

**Figure 4-17 - Typical Lead Von Mises Stress Factor Due to Axial Deflection (psi/mil)**

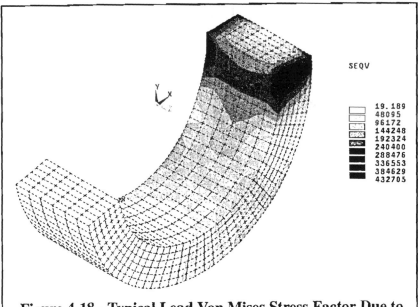

**Figure 4-18 - Typical Lead Von Mises Stress Factor Due to Transverse Deflection (psi/mil)**

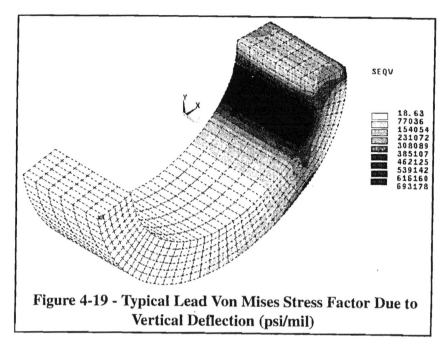

SEQV

|  | 18.63 |
|--|--|
|  | 77036 |
|  | 154054 |
|  | 231072 |
|  | 308089 |
|  | 385107 |
|  | 462125 |
|  | 539142 |
|  | 616160 |
|  | 693178 |

**Figure 4-19 - Typical Lead Von Mises Stress Factor Due to
Vertical Deflection (psi/mil)**

## 4.7.2    Typical Module CTE Determination

Thermal cycling life analysis (see 5.2.3) of surface-mount technology (SMT) usually requires the coefficient of thermal expansion (CTE) on the module surface to be determined. If a complaint bond material is used, this expansion rate can vary over the module surface. Consider a surface-mount module with glass/epoxy PWBs, an aluminum heat sink, and a silicone rubber bond as shown in Figure 4-20. Determine the in-plane CTE in each direction over the surface of the module.

### Development of Finite-Element Model

A finite-element model is generated for the module with 9,464 nodes, 1,550 elements for each PWB, 1,750 elements the heat sink, and 1,550 elements for each bond layer (see Figure 4-21). Material properties are as shown in Table 4-3.

### CTE Determination

The CTE on the surface of the module is determined by first applying a 1°C temperature change to the module finite-element model and determining the thermal strains on the surface of the PWBs. The overall CTE is then determined by adding the CTE of the PWB to the thermal strains for the 1°C temperature change (see Figures 4-22 and 4-23).

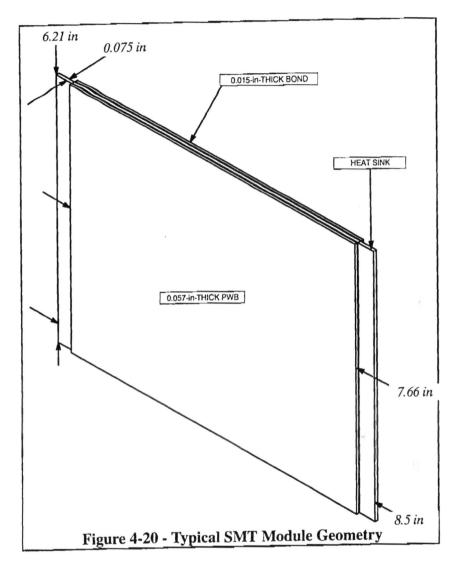

6.21 in

0.075 in

0.015-in-THICK BOND

HEAT SINK

0.057-in-THICK PWB

7.66 in

8.5 in

**Figure 4-20 - Typical SMT Module Geometry**

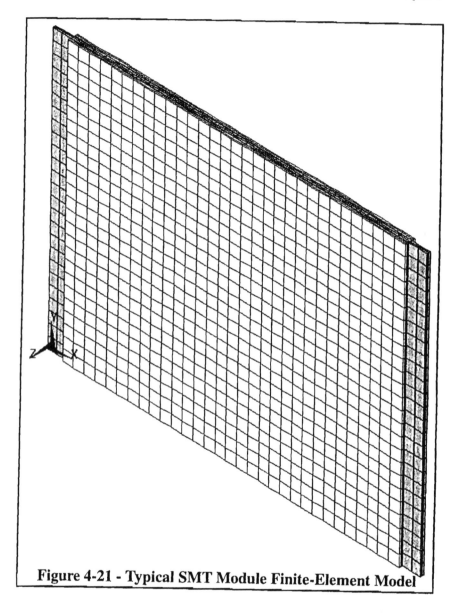

**Figure 4-21 - Typical SMT Module Finite-Element Model**

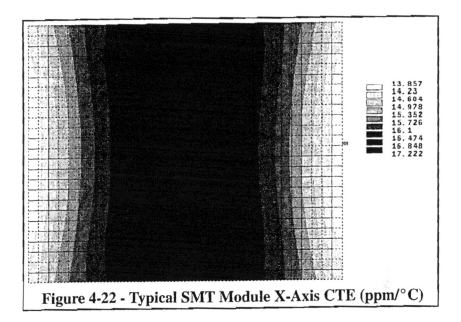

**Figure 4-22 - Typical SMT Module X-Axis CTE (ppm/°C)**

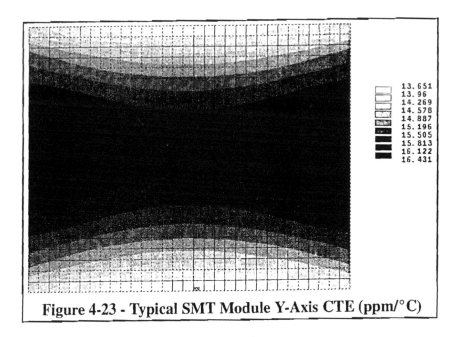

**Figure 4-23 - Typical SMT Module Y-Axis CTE (ppm/°C)**

### 4.7.3    Typical Module Vibration Analysis

Vibration life analysis (see 5.2.2) of SMT usually requires the vibration strains and curvatures on the module surface to be determined. Consider the SMT module described in 4.7.2 (see Figure 4-20). If the the total mass of the module (including components) is 1.16 lbm (see Table 4-4), the overall damping ratio is 10%, and the module is simply supported on three edges. Determine the following for a random vibration input of 0.04 $g^2$/Hz from 50 Hz to 1,000 Hz:

- Module primary resonant frequency

- Module deflection (RMS)

- PWB surface strains

- PWB curvatures

**Development of Finite-Element Model**

The same finite-element model described in 4.7.2 is used for the vibration analysis (see Figure 4-21). Simply supported constraints are applied along both 6.21-in edges and one of the 8.5-in edges. Material properties are as shown in Table 4-3 except the density of the PWB is increased from 0.070 lbm/in³ to 0.132 lbm/in³ to represent the mass of the components "smeared" over the PWB.

### Table 4-4 - Module Mass Distribution

| Location | Volume (in³) | Density (lbm/in³) | Mass (lbm) |
|---|---|---|---|
| Heat Sink | 3.96 | 0.098 | 0.388 |
| 2 Bond layers | 1.43 | 0.040 | 0.057 |
| 2 Glass/epoxy PWBs | 5.42 | 0.070 | 0.379 |
| Components | - | - | 0.336 |
| | | **TOTAL MASS:** | **1.16** |

**Primary Resonant Frequency**

The resonant frequencies are determined by a finite-element modal analysis. The lowest (primary) resonance is found to be *107 Hz* (see Figure 4-24).

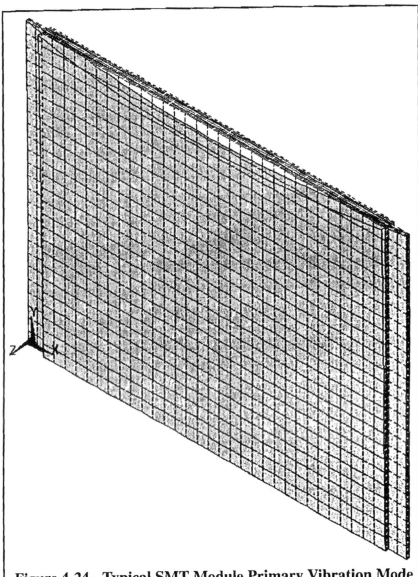

**Figure 4-24 - Typical SMT Module Primary Vibration Mode**

**Module Deflection and Surface Strains**

The module deflection and surface strains are determined by a finite-element spectrum analysis using a 0.04 g2/Hz input. Deflection contours are shown in Figure 4-25, and surface strain contours are shown in Figures 4-26 and 4-27. The deflection spectral density at the center of the module is shown in Figure 4-28.

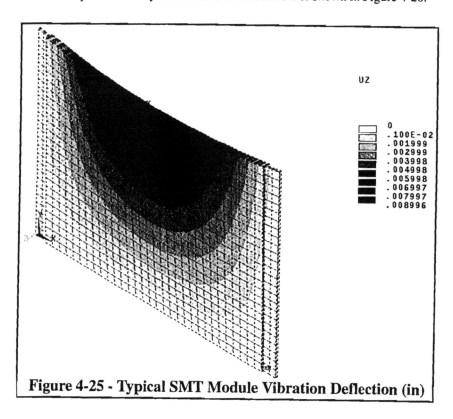

**Figure 4-25 - Typical SMT Module Vibration Deflection (in)**

**Determination of Surface Curvatures**

The surface strains on each side of the PWB are combined with the PWB thickness to determine the radius of curvature for each vibration mode in the frequency range of interest from the finite-element analysis. The curvatures for each mode are then combined with the modal covariance matrix (see equation 4-68) to determine the RMS curvature in each direction (see Figures 4-29 and 4-30). It is important to note that X-axis curvature is due to X-axis strain and that curvature is not expressed as curvature about an axis.

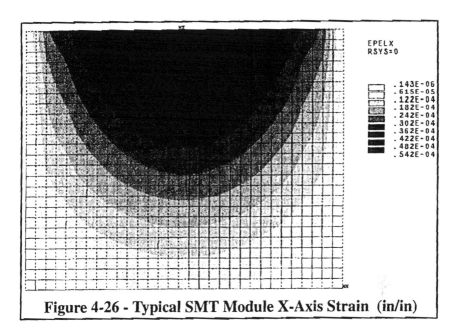

**Figure 4-26 - Typical SMT Module X-Axis Strain  (in/in)**

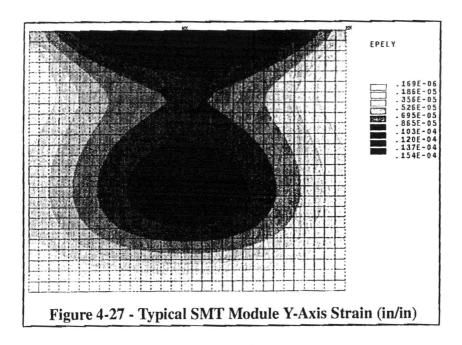

**Figure 4-27 - Typical SMT Module Y-Axis Strain (in/in)**

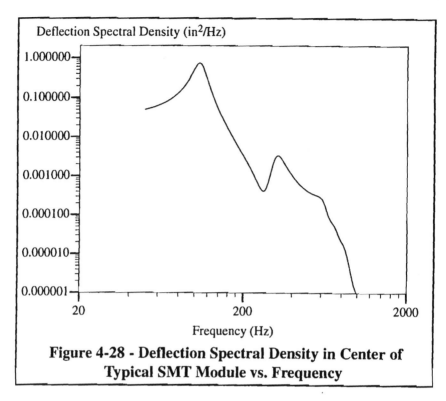

**Figure 4-28 - Deflection Spectral Density in Center of Typical SMT Module vs. Frequency**

### 4.7.4    Typical Acoustic Noise Analysis

Consider a flat aluminum panel as part of an electronic chassis 10.5 in x 6.39 in x 0.0625 in thick. This panel is exposed to an acoustic input of 140 db from 100 Hz to 2,000 Hz based upon one-third octave analysis (see Figure 4-10). Consider the damping to be 1.5%. Determine:

- Maximum (RMS) stress in panel
- Maximum (RMS) deflection of panel
- Maximum acoustic intensity inside chassis

**Development of Finite-Element Model**

A finite-element model is generated using 504 shell elements and 551 nodes with six-degrees of freedom at each node (see Figure 4-31). Simply supported constraints are applied at the boundaries of the panel to represent the worst-case (least stiffness) condition. Material properties are as shown in Table 4-3.

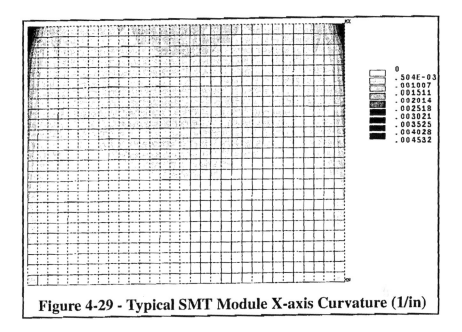

**Figure 4-29 - Typical SMT Module X-axis Curvature (1/in)**

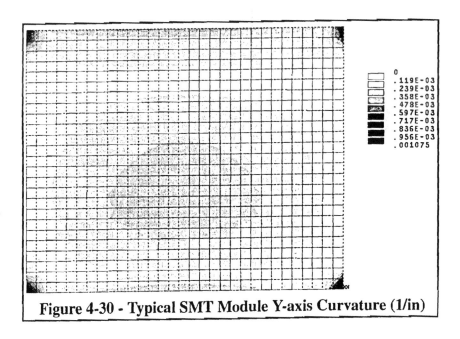

**Figure 4-30 - Typical SMT Module Y-axis Curvature (1/in)**

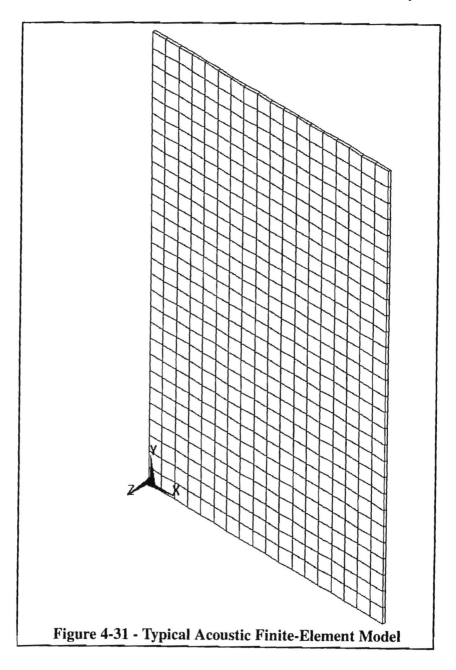

**Figure 4-31 - Typical Acoustic Finite-Element Model**

### Harmonic Response for Constant Input Pressure

A harmonic response finite-element analysis is conducted using a harmonically varying pressure load of 1 psi, producing the maximum (center of panel) Von Mises stress vs. frequency and deflection vs. frequency curves shown in Figures 4-32 and 4-33.

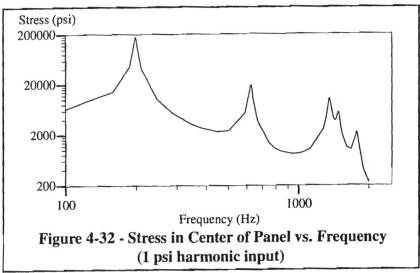

**Figure 4-32 - Stress in Center of Panel vs. Frequency
(1 psi harmonic input)**

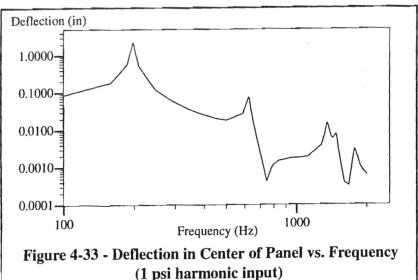

**Figure 4-33 - Deflection in Center of Panel vs. Frequency
(1 psi harmonic input)**

**Acoustic Response**

The acoustic response to the input environment is determined by first converting the input acoustic spectrum to a pressure spectral density:

$PBL$ = *Pressure Band Level* = 140 *db*
$M$ = *Slope of Band SPL* = 0 *dB/octave*
$K$ = *Band Width Factor* = 0.2316 *(for 3 bands/otave from Table 4-2)*
$p_0$ = *Reference Pressure* = 0.0002 *μbar* = 2.95 x $10^{-9}$ *psi (in air)*

$$A = PBL_1 - M \frac{\log(f_1)}{\log(2)}$$

$$A = 140 \, db - 0 = 140 \, db \tag{4-37}$$

$$B = \frac{M}{\log(2)}$$

$$B = \frac{0}{\log(2)} = 0 \tag{4-36}$$

$$PBL = A + B \, \log(f)$$

$$PBL = 140 \, db + 0 \, \log(f) = 140 \, db \tag{4-35}$$

$$PrSD = \frac{p_0^2 \, 10^{\left(\frac{A}{10}\right)}}{K} f^{\left(\frac{B}{10}-1\right)} = \frac{(2.95 \times 10^{-9} \, psi)^2 \, 10^{\left(\frac{140db}{10}\right)}}{0.2316} f^{\left(\frac{0}{10}-1\right)} \tag{4-43}$$

$$PrSD = \frac{3.76 \times 10^{-3}}{f} \frac{psi^2}{Hz}$$

$f$ = *Frequency (Hz)*

Using this pressure spectral density in conjunction with the width of each frequency band (from the harmonic analysis) to determine the input pressure at each frequency. This input pressure is then used to scale the harmonic response stresses and/or deflections which are then RSS summed over the frequency range to determine overall RMS response (see Figures 4-34 and 4-35). Alternatively, the above pressure spectral density can be multiplied by the square of the deflection and stress responses shown in Figures 4-32 and 4-33 to determine the deflection and stress spectral densities (see Figures 4-36 and 4-37). Maximum RMS stress and deflection of *2,703 psi RMS* and *0.0334 in RMS* is obtained directly from Figures 4-34 and 4-35 or by taking the square root of the area under Figures 4-36 and 4-37.

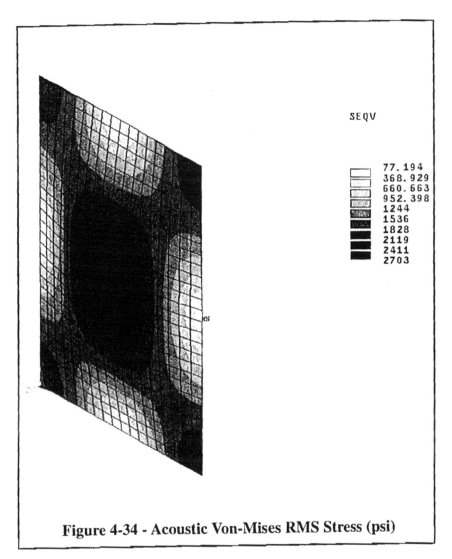

SEQV

77.194
368.929
660.663
952.398
1244
1536
1828
2119
2411
2703

**Figure 4-34 - Acoustic Von-Mises RMS Stress (psi)**

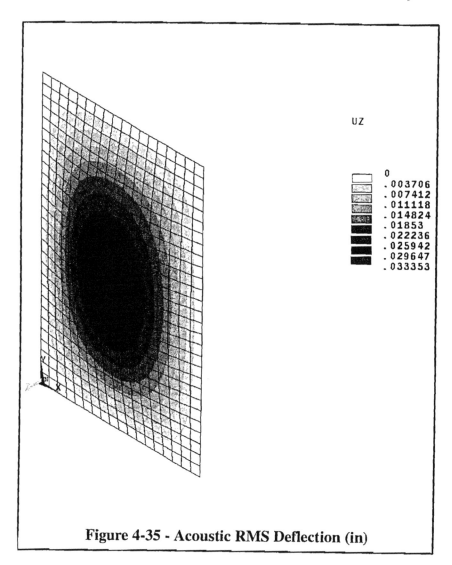

**Figure 4-35 - Acoustic RMS Deflection (in)**

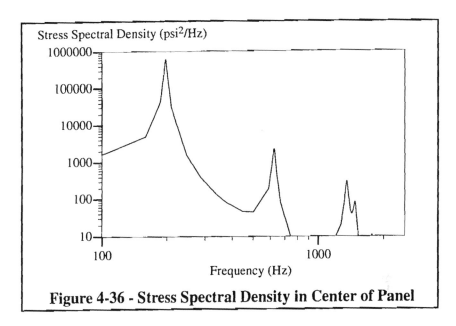

**Figure 4-36 - Stress Spectral Density in Center of Panel**

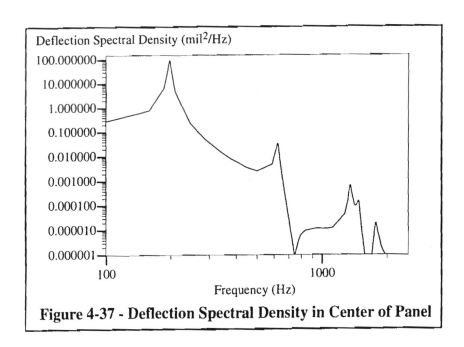

**Figure 4-37 - Deflection Spectral Density in Center of Panel**

**Acoustic Intensity Inside Chassis due to Wall Motion**

The deflection of the panel is used to determine the pressure amplitude from the wall motion as follows:

$f$ = *Frequency of Oscillation (peak response)* = 198 *Hz*

$\rho_0 c$ = *Specific Acoustic Impedance (air at 20ºC)* = $415\dfrac{kg}{m^2\ sec}$[25]

$\delta$ = *Wall Deflection* = 0.0334 *in (RMS)*

$$p = \rho_0 c\ 2\pi f\delta = 415\frac{kg}{m^2\ sec}\ 2\pi\ x\ 198\ Hz\ x\ 0.0334\ in\ x\frac{m}{39.37in}$$

(4-47)

$$p = 438\frac{N}{m^2}x\frac{lb}{4.448\ N}\left(\frac{m}{39.37\ in}\right)^2 = 0.0635\ psi\ RMS$$

Alteratively, equation 4-47 may be solved for the pressure to deflection ratio as follows:

$$\frac{p}{\delta} = \rho_0 c\ 2\pi f = 415\frac{kg}{m^2\ sec}\ 2\pi f = (2608\ f\ )\frac{kg}{m^2\ sec^2}\ (using\ f\ in\ Hz)$$

$$\frac{p}{\delta} = 2608\ f\frac{kg}{m^2\ sec^2} = (2608\ f\ )\frac{N}{m^2\text{-}m}x\frac{lb}{4.448\ N}\left(\frac{m}{39.37\ in}\right)^3$$ (4-80)

$$\frac{p}{\delta} = (9.608x10^{-3}\ f\ )\ \frac{psi}{in}\ (using\ f\ in\ Hz)$$

The square of the above ratio, multiplied by the deflection spectral density, gives the pressure spectral density inside the chassis near the center of the panel (see Figure 4-38).

**Acoustic Intensity Inside Chassis due to Transmission**

Pressure spectral density inside the chassis due to direct transmission through the wall is calculated as follows:

$l$ = *Wall Thickness* = 0.0625 *in*

$\rho_a c_a$ = *Specific Acoustic Impedance of Air (at 20ºC)* = $415\dfrac{kg}{m^2\ sec}$  [25]

$\rho_w c_w$ = *Specific Acoustic Impedance of Aluminum* = $13.9x10^6\dfrac{kg}{m^2\ sec}$  [25]

$c_w$ = *Speed of Sound in Aluminum* = $5,150\dfrac{m}{sec}$  [25]

$$k_w = \frac{\omega}{c_w} = \frac{2\pi f}{c_w} = (1.22x10^{-3} f )\frac{1}{m} \quad (using\, f\, in\, Hz.) \qquad (4\text{-}49)$$

$$a_t = \frac{4(\rho_a c_a)^2}{(\rho_w c_w)^2 (k_w l)^2}$$

$$a_t = \frac{4\left(415\frac{kg}{m^2\, sec}\right)^2}{\left(13.9x10^6 \frac{kg}{m^2\, sec}\right)^2 \left((1.22x10^{-3} f )\frac{1}{m} \, x\, \frac{m}{39.37\, in}\, x\, 0.0625\, in\right)^2} \qquad (4\text{-}48)$$

$$a_t = \frac{950.6}{f^2} \quad (using\, f\, in\, Hz)$$

$$PrSD_t = a_t\, PrSD_i = \frac{950.6}{f^2}\, PrSD_i\, (using\, f\, in\, Hz) \qquad (4\text{-}52)$$

High relative impedance condition check:

$$\rho_w c_w \gg \rho_a c_a$$

$$13.9x10^6 \frac{kg}{m^2\, sec} \gg 415\frac{kg}{m^2\, sec}$$

Thin wall condition check:

$$k_w l \ll 1$$

$$k_w = (1.22x10^{-3} \, x\, 2{,}000\, )\frac{1}{m} = 2.44\frac{1}{m} \quad (at\, 2{,}000\, Hz.)$$

$$\left(2.44\frac{1}{m}\, x\, \frac{m}{39.37\, in}\, x\, 0.0625 in = 0.0039\right) \ll 1$$

The pressure spectral density inside the chassis due to direct transmission is included in Figure 4-38.

**Total Acoustic Intensity Inside Chassis**

The total acoustic intensity inside the chassis is determined by first calculating the total pressure spectral density inside the chassis as the sum of the pressure spectral densities due to wall motion and transmission (see Figure 4-38). Taking the square root of the area under the total pressure spectral density response gives the total RMS pressure amplitude inside the chassis of 0.065 psi. The RMS pressure amplitude is converted to decibels to determine the overall SPL as follows:

$p$ = *Pressure Amplitude* = 0.065 *psi*

$p_0$ = *Reference Pressure* = 0.0002 *μbar* = 2.95 x $10^{-9}$ *psi (in air)*

$$SPL = 20\log\left(\frac{p}{p_0}\right) = 20\log\left(\frac{0.065psi}{2.95x10^{-9}\,psi}\right) = \mathbf{146.9\,db} \qquad (4\text{-}33)$$

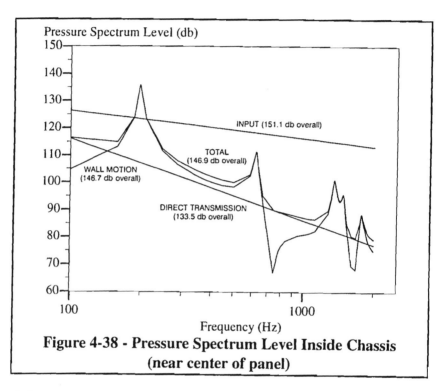

**Figure 4-38 - Pressure Spectrum Level Inside Chassis (near center of panel)**

## 4.8 INTERPRETATION OF RESULTS

Results of mechanical analysis must be properly interpreted to provide the desired results. Typically, finite-element software calculates the direct and shear stresses components in each of the coordinate directions. If these results are to be compared to a yield stress or other material capability, some method of combining the stress components is required. For brittle materials, the maximum (tensile) principal stress is used. Principal stresses represent the stresses in three orthogonal directions (maximum, intermediate, minimum) in which the shear stress is zero. The eigenvalues of the stress component matrix determine the principal stresses:

$$[\sigma] = \begin{bmatrix} \sigma_x & \tau_{xy} & \tau_{xz} \\ \tau_{xy} & \sigma_y & \tau_{yz} \\ \tau_{xz} & \tau_{yz} & \sigma_z \end{bmatrix} \tag{4-81}$$

Eigenvalues of the above matrix are given by the roots of the following cubic equation:

$$\begin{aligned} &\sigma_i^3 - (\sigma_x + \sigma_y + \sigma_z)\,\sigma_i^2 + \\ &(\sigma_x\sigma_y + \sigma_x\sigma_z + \sigma_y\sigma_z - \tau_{xy}^2 - \tau_{xz}^2 - \tau_{yz}^2)\,\sigma_i + \\ &(\tau_{xy}^2\sigma_z + \tau_{xz}^2\sigma_y + \tau_{yz}^2\sigma_x - \sigma_x\sigma_y\sigma_z - 2\tau_{xy}\tau_{xz}\tau_{yz}) = 0 \end{aligned} \tag{4-82}$$

$[\sigma]$ = *Stress Component Matrix*
$\sigma_{x,y,z}$ = *Direct Stress Components in x,y,z Directions*
$\tau_{xy,yz,xz}$ = *Shear Stress Components in xy,yz,xz Directions*
$\sigma_i$ = *Principal Stresses*

For ductile materials, the Von Mises stress is usually used:

$$\sigma_e = \sqrt{\frac{(\sigma_1 - \sigma_2)^2 + (\sigma_2 - \sigma_3)^2 + (\sigma_3 - \sigma_1)^2}{2}} \tag{4-83}$$

$\sigma_e$ = *Equivalent (Von Mises) Stress*
$\sigma_{1,2,3}$ = *Principal Stresses*

It is important to apply the appropriate factors of safety when comparing the developed stresses described above to material capabilities. In random vibration and acoustic analyses, it is also important to consider the peak stresses as well as the RMS stresses as appropriate (see Table 4-1).

For thermal cycling analysis of solder joints (see 5.2), the effective CTE (in each direction) of the module surface is required. Vibration analysis of solder joints (see 5.2.2) requires the dynamic strain and curvature components on the module surface. These guidelines are summarized in Table 4-5.

| Table 4-5 - Results Interpretation Summary | | | |
|---|---|---|---|
| **Type of Analysis** | **Material** | **Result Used** | **Approach** |
| Static Mechanical | Ductile | Von Mises Stress | Compare to yield stress* |
| | Brittle | Maximum Principal Stress | Compare to tensile stress* |
| Dynamic Mechanical | Ductile | Von Mises Stress | Use in life analysis (see Chapter 5) |
| | Brittle | Maximum Principal Stress | |
| | Solder | Module Surface Strains and Curvature | |
| Thermal Cycling | Solder | Module Surface CTE | Use in solder-life analysis (see 5.2.3) |
| | | | * - With appropriate factors of safety |

# 4.9   FACTORS IN RECENT DEVELOPMENTS

The principles described in this chapter are applicable to a wide range of electronic packaging configurations. Mechanical analysis of recent electronic packaging configurations (see 1.1.3) would typically require component-level finite-element analysis (see 4.3.2). Of course, guidelines and assumptions described for the appropriate conventional electronic packaging analyses apply.

**Ball-Grid Arrays (see Figure 1-5)**

A finite-element model of a BGA would typically include the package, the solder balls, and a portion of the PWB [31]. If an underfill is used, it should be included in the model.

**Multichip Modules (see Figure 1-6)**

Since the chips in a multichip module are lightweight and securely attached to a rigid substrate, analysis for typical dynamic mechanical environments may not be required unless the environment is unusually severe. However, if the expansion rates of the individual parts are different; a thermal-stress analysis would be required (see 4.6.3).

### Flip-Chips (see Figure 1-7)

Flip-chips are also typically lightweight and securely attached to a stiff substrate, so analysis for dynamic mechanical environments may not be required unless the environment is severe. Once again, if the expansion rates of the substrate and die are different; a thermal-stress analysis would be required (see 4.6.3 and [32]).

### Chip-on-Board (see Figure 1-8)

Chip-on-board (COB) configurations, although typically lightweight and securely attached to the substrate, may require analysis for dynamic mechanical environments due to the dynamic response of the PWB under the COB. Because the level of detail for a chip-on-board model is greater than that required for a PWB/module analysis, substructuring may be helpful (see 4.3.1).

Since the expansion rates of the various parts are usually different, a thermal-stress analysis is typically required (see 4.6.3). It is important that the analysis consider the effect of the encapsulant, especially for the encapsulant that is between the die and PWB in flip-chip COB applications.

## 4.10  VERIFICATION

Results from mechanical analysis should be verified to help avoid any inaccuracies that might arise. The verification method may be testing, use of a simplified model, altered mesh density, or other methods as appropriate. For information on verification, see Chapter 8.

## 4.11  MECHANICAL ANALYSIS CHECKLIST

### 4.11.1  Applicable to All Mechanical Analyses

☐ Has the input environment been properly defined?

☐ Have appropriate boundary conditions been applied at all levels of the analysis?

☐ Does the mass used in the analysis agree with the appropriate system weight?

☐ Have appropriate factors of safety been applied to results?

☐ Has all source material (material properties, environmental data, etc.) been verified against the references?

### 4.11.2    Applicable to Finite-Element Analysis

☐  Does the element selection match the expected loading?

☐  Is the mesh density consistent with the expected loading?

☐  Are degrees of freedom consistent where dissimilar elements connect?

☐  Has a check been made for nodes at the same location but connected to different elements?

☐  Are material types properly assigned?

### 4.11.3    Applicable to Dynamic Analyses

☐  Has damping been properly characterized?

☐  Has proper distinction been made between absolute and relative accelerations, displacements, etc.?

☐  Does the analysis cover the full frequency range of the input environment?

☐  Has proper consideration been made to the relationship between peak and RMS results for random vibration and acoustic analyses?

### 4.11.4    Applicable to Acoustic Analysis

☐  Has proper consideration been given to pressure band level, pressure spectrum level, and overall sound pressure level?

☐  Has the analysis bandwidth (third octave, etc.) been considered?

### 4.11.5    Applicable to Recent Developments

☐  Is any underfill or encapsulant included in the model?

### 4.11.6    Environmental Data Required

The following environmental data is typically required for various analyses. Typical units are provided for convenience, and care must be taken to ensure that the system of units is consistent with the material properties and model dimensions.

**Sinusoidal Vibration Analysis**

☐  Vibration amplitude (g, cm, cm/sec, in, etc.)

☐  Vibration frequency (Hz)

☐  Vibration duration (sec, hr, etc.)

☐  Vibration direction

**Random Vibration Analysis**

- [ ] Power spectral density ($g^2$/Hz) vs. frequency (Hz) curve
- [ ] Vibration duration (sec, hr, etc.)
- [ ] Vibration direction

**Mechanical Shock Analysis**

- [ ] Shock amplitude (g) vs. time (sec)
- [ ] Number of shocks
- [ ] Shock direction

**Acoustic Analysis**

- [ ] Pressure spectrum level (db) or pressure band level (db) vs. frequency (Hz)
- [ ] Duration of acoustic input (sec, hr, etc.)

**Sustained Acceleration Analysis**

- [ ] Acceleration amplitude (g)
- [ ] Acceleration direction

**Thermal Shock Analysis**

- [ ] Initial temperature (°C, °F, etc.)
- [ ] Final temperature (°C, °F, etc.)
- [ ] Ramp rate (°C/sec, °F/sec, etc.)

## 4.11.7  Material Properties Required

The following material properties are typically required for various analyses. Typical units are provided for convenience, and care must be taken to ensure that the system of units is consistent for any analysis. If materials are not isotopic, components of various properties in each direction may be required.

**Static Mechanical Analysis**

Two of the following:

- [ ] Modulus of elasticity (MPa, psi, etc.)
- [ ] Poisson's ratio (dimensionless)
- [ ] Shear modulus (MPa, psi, etc.)

Sustained acceleration static analysis also requires:

☐   Density (kg/m$^3$, slug/ft$^3$, etc.)

**Dynamic Mechanical Analysis**

☐   Material properties from above

☐   Density (kg/m$^3$, slug/ft$^3$, etc.)

☐   Material-dependent damping ratio (dimensionless)

**Acoustic Analysis**

☐   Material properties from above

☐   Acoustic reference pressure in the surrounding medium (μbar, MPa, psi, etc.)

It also requires one of the following:

☐   Speed of sound in material (m/sec, ft/sec, etc.)

☐   Acoustic impedance (kg/(m$^2$-sec), rayl, etc.)

**Thermal Shock Analysis**

☐   Material properties required for static analysis (see above)

☐   Coefficient of thermal expansion (ppm/°C, in/in/°C, in/in/°F, etc.)

☐   Thermal conductivity (W/(m-°C), Btu/(hr-ft-°F), etc.)

☐   Specific heat (J/(kg-°C), cal/(g-°C), Btu/(lbm-°F), etc.)

☐   Density (kg/m$^3$, slug/ft$^3$, etc)

## 4.12  REFERENCES

[20]   Noble, B.; Applied Linear Algebra; Prentice-Hall, 1969

[21]   McKeown, S.; "Automated Solder Life Analysis of Electronic Modules", (presented at the 126th TMS Annual Meeting and Exhibition, Orlando, FL, February 10, 1997); Design and Reliability of Solders and Solder Interconnections; The Minerals, Metals & Materials Society, Warrendale, PA, 1997; pp 279-286

[22]   Kohnke, P.; ANSYS Theory Reference, Release 5.3, Seventh Edition; ANSYS Incorporated, 201 Johnson Road, Houston, PA; 1996

[23] ANSYS Advanced Analysis Techniques Guide, Release 5.3, First Edition; ANSYS Incorporated, 201 Johnson Road, Houston, PA; June 1996

[24] Dimarogonas, A.; Vibration Engineering; West Publishing Co., 1976

[25] Kinsler, L. and Frey, A.; Fundamentals of Acoustics, Second Edition; John Wiley and Sons, 1963

[26] Seto, W.; Schaum's Outline of Theory and Problems of Acoustics; McGraw-Hill, 1971

[27] Materials Engineering, 1992 Materials Selector; Penton Publishing, Cleveland, OH; December 1991

[28] McClintock, F.A. and Argon, A.S.; Mechanical Behavior of Materials; Addison-Wesley, 1966

[29] Solomon, Brzozowski, Thompson; "Prediction of Solder Joint Fatigue Life", Report Number 88CRD01; General Electric Corporate Research and Development, Schenectady, NY, 1988

[30] Dance, F.J.; "Mounting Leadless Ceramic Chip Carriers Directly to Printed Wiring Boards - A Technology Review and Update"; Circuits Manufacturing, May 1983

[31] Cheng, Chiang, Lee; "An Effective Approach for Three-Dimensional Finite Element Analysis of Ball Grid Array Typed Packages"; Journal of Electronic Packaging, Volume 120, Number 2; American Society of Mechanical Engineers, June 1998; pages 129-134

[32] Wang, Qian, Zou, Lou; "Creep Behavior of a Flip-Chip Package by both FEM Modeling and Real Time Moiré Interferometry"; Journal of Electronic Packaging, Volume 120, Number 2; American Society of Mechanical Engineers, June 1998; pages 179-185

# Chapter 5
# Life Analysis

## 5.1  BACKGROUND

The mechanical analysis techniques described in Chapter 4 explained the calculation of the stress and strain due to dynamic and static loads. Although in some cases the developed stress and strain could be compared to static material properties (see 4.8), usually fatigue is the governing factor in electronic packaging analysis. Fatigue is the "failure of a material by cracking resulting from repeated or cyclic stress" [48]. When the fatigue of the materials is compared to the cyclic loads expected in the operational, storage, and test environments, the expected life can be determined. Sources of these cyclic loads include the following:

- Vibration

- Repeated mechanical shock

- Acoustic noise

- Thermal cycling

- Temperature extremes

For electronic packaging configurations, much of the life analysis activity is concentrated on interconnections (solder joints, leads, plated-through holes, etc.), although concern must also be directed to chassis and other structural areas.

## 5.1.1  High-Cycle Fatigue

Fatigue life under dynamic mechanical (vibration, shock, acoustic, etc.) loads is typically governed by high-cycle fatigue. High-cycle fatigue is typically predicted by a traditional S-N (stress vs. cycles) curve based upon the applied stress. To meet the definition of high-cycle fatigue, the applied stress must be less than the yield stress so that no plastic deformation is present. In many cases, the S-N curve is a straight line on log-log axes, so the following equation can be written for fatigue life:

$$n_f = \left(\frac{K'}{\sigma}\right)^{\lambda} \qquad\qquad (5\text{-}1)$$

$n_f$ = Mean Cycles to Fail
$K'$ = Fatigue Life Multiplier
$\sigma$ = Peak Stress
$\lambda$ = Fatigue Exponent

## 5.1.2    Low-Cycle Fatigue

If yielding is present in a material, fatigue life is governed by low-cycle fatigue. Low-cycle fatigue is typically predicted by the Manson-Coffin relationship [33]:

$$n_f = \frac{1}{2}\left(\frac{\Delta\epsilon_p}{2\epsilon_f'}\right)^{\frac{1}{c}} \qquad\qquad (5\text{-}2)$$

$n_f$ = Mean Cycles to Fail
$\Delta\epsilon_p$ = Plastic Strain
$2\epsilon_f'$ = Fatigue Ductility Coefficient
$c$ = Fatigue Ductility Exponent

## 5.1.3    Random Vibration

**Random Vibration Cycles to Fail**

Random vibration (see 4.4.3) produces a normal distribution of amplitudes (see Table 4-1). If the high-cycle fatigue equation (5-1) is combined with the cumulative damage equation (5-14) using the distribution of amplitudes given in Table 4-1, a fatigue equation for random vibration can be derived:

$$1 = \frac{0.6826\, n_f}{\left(\frac{K'}{\sigma}\right)^\lambda} + \frac{0.2718\, n_f}{\left(\frac{K'}{2\sigma}\right)^\lambda} + \frac{0.0456\, n_f}{\left(\frac{K'}{3\sigma}\right)^\lambda} \qquad (5\text{-}3)$$

$$1 = \frac{0.6826\, n_f\, \sigma^\lambda}{(K')^\lambda} + \frac{0.2718\, n_f\, (2\sigma)^\lambda}{(K')^\lambda} + \frac{0.0456\, n_f\, (3\sigma)^\lambda}{(K')^\lambda} \qquad (5\text{-}4)$$

$$n_f = \frac{(K')^\lambda}{(0.6826 + 0.2718\,(2)^\lambda + 0.0456\,(3)^\lambda)\,\sigma^\lambda} \qquad (5\text{-}5)$$

$$n_f = \frac{(K')^\lambda}{(C\,\sigma)^\lambda} \qquad (5\text{-}6)$$

$n_f$ = Mean Cycles to Fail
$K'$ = Fatigue Life Multiplier
$\sigma$ = RMS Stress
$\lambda$ = Fatigue Exponent
$C$ = Random Vibration Amplitude Factor (see Table 5-1)

The random vibration amplitude factor (C) in equation 5-6 can be used to convert a RMS random stress amplitude to an equivalent peak sinusoidal vibration amplitude. This factor depends only upon the fatigue exponent ($\lambda$) and is shown for various exponents in Table 5-1.

| Table 5-1 - Random Vibration Amplitude Factor | |
|---|---|
| Fatigue Exponent ($\lambda$) | Amplitude Factor (C) |
| 2 | 1.48 |
| 4 | 1.72 |
| 6 | 1.93 |
| 8 | 2.09 |
| 10 | 2.22 |
| 12 | 2.33 |
| 16 | 2.47 |

**Random Vibration Cycle Counting**

One of the concerns that arises with life analysis of random signals is the method for counting the cycles. Since the signal is random, and a large number of frequencies are present at the same time (see 4.4.3), it is not immediately obvious how to count the cycles to determine life. If resonances are present in the item under consideration, the stresses will peak at the resonant frequencies so they may be used to relate the number of cycles to the duration of the random signal. Since the stresses in electronic packaging configurations are driven by displacement, use of the lowest resonant frequency usually produces good results. It is important, however, to verify (by plotting stress vs. frequency, etc.) that the correct frequency is being used for cycle counting. The relationship between cycles and time to failure is thus expressed as follows:

$$t_f = \frac{n_f}{f_n}$$  (5-7)

$n_f$ = Mean Cycles to Fail

$t_f$ = Time to Fail

$f_n$ = Primary (most significant) Resonant Frequency

If stresses do not have a dominant frequency, a time domain method such as the rain flow algorithm [45] may be required.

## 5.1.4    Effect of Curvature on Out-of-Plane Deflection

Vibration of printed wiring boards (PWBs) typically results in out-of-plane deflection which contributes to the stress in the solder and leads for components with more than 3 terminals. The relationship between curvature and out-of-plane deflection is given by the intersecting chords theorem (see Figure 5-1) as follows:

$$(2R - \delta)\,\delta = \left(\frac{L}{2}\right)^2$$  (5-8)

$$(2R\delta - \delta^2) = \left(\frac{L}{2}\right)^2$$  (5-9)

If $\delta$ is small compared to $R$:

$$2R\delta \approx \left(\frac{L}{2}\right)^2$$  (5-10)

$$\delta \approx \left(\frac{L}{2}\right)^2 \frac{1}{2R}$$  (5-11)

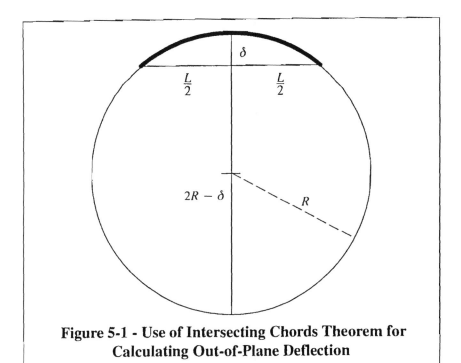

**Figure 5-1 - Use of Intersecting Chords Theorem for
Calculating Out-of-Plane Deflection**

Out-of-plane deflection of a component terminal is half of δ since the component
will find a neutral point between the high and low points:

$$\delta_{oop} = \frac{\delta}{2} = \frac{L^2}{16}\mathcal{C} \qquad (5-12)$$

$$\mathcal{C} = \frac{1}{R} \qquad (5-13)$$

$R$ = *Radius of Curvature*
$\delta$ = *Total Deflection*
$L$ = *Component Length*
$\delta_{oop}$ = *Out-of-plane Deflection of Component Terminal*
$\mathcal{C}$ = *Curvature (see 4.5.3)*

## 5.1.5   Cumulative Damage

If cyclic loads are applied at various levels, failure is thought to occur when the
total cumulative damage is unity. Cumulative damage is the sum of ratio between

the cycles or time applied at a given level, to the cycles or time to fail at that level. Cycles and time to fail may be used interchangeably, as follows:

$$D = \sum \left( \frac{n_i}{n_{f_i}} \text{ or } \frac{t_i}{t_{f_i}} \right)$$                (5-14)

$D = $ Cumulative Damage
$n_{f_i} = $ Cycles to Fail at Level i
$n_i = $ Cycles Applied at Level i
$t_{f_i} = $ Time to Fail at Level i
$t_i = $ Duration at Level i

The time or cycles to fail may be the mean cycles to fail or, if the failure distribution is known, the cycles to a pre-determined cumulative failure percentage.

### 5.1.6    Combined Tensile and Alternating Stresses

When a material is subjected to a combined tensile and alternating stress, the fatigue capacity is reduced. Two methods of accounting for this phenomenon are the Goodman and Soderberg diagrams in which the alternating stress ($\sigma_a$) is plotted on the vertical axis and the tensile stress ($\sigma_m$) is plotted on the horizontal axis (see Figure 5-2) [47]. In the Goodman diagram, a straight line is plotted between the endurance stress ($S_e$, maximum stress for infinite fatigue life) for a material and the ultimate stress ($S_u$). If the current stress state ($\sigma_m, \sigma_a$) is below and to the left of the Goodman line, no fatigue failure is expected. The Soderberg diagram is similar but more conservative since the yield stress ($S_y$) is used instead of the ultimate stress ($S_u$). For materials that do not have an endurance limit or in situations where infinite life is not required, a modified Soderberg line may be plotted by using the stress for N cycles to fail ($S_N$) in place of the endurance stress ($S_e$). Using the slope of the modified Soderberg line, an equivalent alternating stress may be developed as follows:

$$\sigma_{a_{eq}} = S_N = \sigma_a \frac{S_y}{S_y - \sigma_m}$$                (5-15)

$\sigma_{a_{eq}} = $ Equivalent Alternating Stress
$S_N = $ Stress for N Cycles to Fail
$\sigma_a = $ Alternating Stress
$S_y = $ Yield Stress
$\sigma_m = $ Tensile (mean) Stress

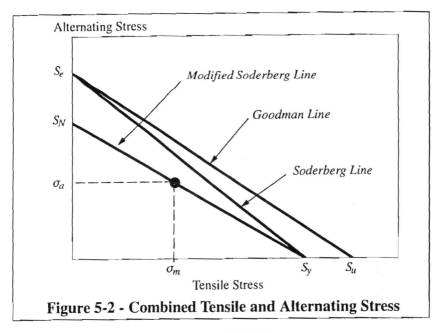

**Figure 5-2 - Combined Tensile and Alternating Stress**

## 5.2   SOLDER-LIFE ANALYSIS

### 5.2.1   Applications

**Surface-Mount Technology**

Configurations considered include surface-mount technology (SMT) modules consisting of two printed wiring boards (PWBs) bonded to a centrally mounted heat sink (see Figures 5-3 and 5-4). Leadless and leaded chip-carriers, and ball-grid arrays (see Figure 5-5) are supported by the solder. The integrity of the electrical connection for these components is influenced by the geometry and fatigue of the solder. SMT may also be used in single-board configurations with components mounted to one or both sides of the PWB, and may also include the use of through-hole components (mixed technology).

Leadless Components

Electronic module configurations using leadless components and ball-grid arrays are more sensitive to vibration and thermal cycling because there is very little compliance between the component and the module surface. This lack of compliance in conjunction with surface strains on the printed wiring board (PWB) due

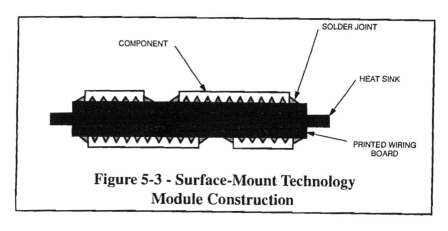

**Figure 5-3 - Surface-Mount Technology
Module Construction**

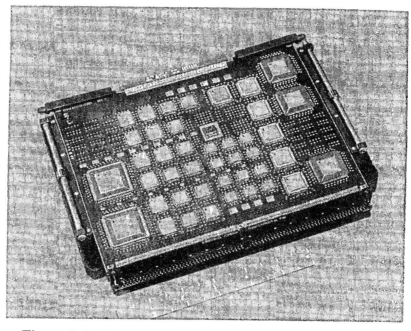

**Figure 5-4 - Typical Surface-Mount Technology Module**
(Courtesy of Lockheed Martin Control Systems.)

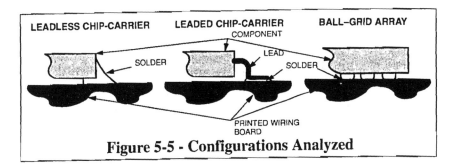

**Figure 5-5 - Configurations Analyzed**

to vibration can result in high vibration stresses which can lead to premature solder failure due to fatigue. Alternatively, this lack of compliance in conjunction with mismatched expansion rates of the module surface and components, and/or the solder and component body can result in high thermal strains which can lead to low-cycle fatigue of the solder.

Leaded Components

Leaded components provide additional compliance which helps to alleviate the vibration stress and thermal strain developed in the solder. Although there is additional compliance, the component is still supported by the solder and is still subject to solder fatigue. In addition, the combination of thermal cycling and vibration may result in lead failure due to combined static and dynamic stresses in the lead.

**Through-Hole Technology**

In through-hole technology configurations, the solder is partially supported by the geometry of the lead in the hole, and the solder serves to enhance the electrical conduction between the lead and the PWB. In many cases the component would be retained on the printed wiring board even if no solder was present. If the lead is primarily supported by the geometry of the assembly, the solder is not exposed to significant stress or strain so prediction of solder joint life is difficult. In such cases a relative life analysis (see 5.5) based upon test data may be appropriate.

## 5.2.2     Solder Joint Life under Vibration

**Approach**

Leadless Chip-Carriers and Ball-Grid Arrays

1.  Vibration-induced surface strain and curvature (see Table 4-5) is used to calculate shear strain in the solder

2. Solder shear strain multiplied by shear modulus is used to determine shear stress in the solder

## Leaded Chip-Carriers

1. Vibration-induced surface strain and curvature (see Table 4-5) is used to calculate lead force
2. Force divided by effective area of solder joint is used to determine shear stress

## Both Leaded and Leadless Chip-Carriers

3. Shear stress is used to calculate solder life
4. Solder life corrected by a factor to represent solder reliability

## Leadless Chip-Carriers and Ball-Grid Array Vibration Stress

The vibration stress in a leadless chip-carrier or ball-grid array solder joint is given by the following relationship:

$$\tau = G \frac{\gamma}{S} \tag{5-16}$$

$$\gamma = \sqrt{\gamma_x^2 + \gamma_y^2 + \epsilon_z^2} \tag{5-17}$$

$$\gamma_{x,y} = \epsilon_{x,y} \frac{L_{x,y}}{2h} \tag{5-18}$$

$$\epsilon_z = \frac{\delta_z}{h} \tag{5-19}$$

$$\delta_z = \max\left( \frac{L_x^2}{16} C_x \,, \, \frac{L_y^2}{16} C_y \right) \tag{5-20}$$

$\tau = $ *Solder Shear Stress*

$G = $ *Solder Shear Modulus (see Table 5-3)*

$\gamma = $ *Maximum Shear Strain*

$S = $ *Strain Factor for Last Zone to Crack (see Table 5-2)*

$\gamma_{x,y} = $ *In-Plane Shear Strain in x,y Direction*

$\epsilon_z = $ *Out-of-Plane Solder Strain*

$\epsilon_{x,y} = $ *PWB Surface Vibration Strain in x,y Direction*

$L_{x,y} = $ *Length of Package Footprint in x,y Direction*

$h = $ *Solder Joint Height*

$\delta_z$ = *Maximum Out-of-Plane Deflection*

$\mathcal{C}_{x,y}$ = *Curvature in x,y Direction (see 4.5.3)*

## Crack Zones

The strain in the castellated leadless chip-carrier solder joint varies from a maximum under the package to a minimum in the fillet. Since the strain is not constant in the chip-carrier solder joint, any analytical method must account for this variation in strain. An excellent technique was described in a paper by Solomon et. al. [37]. In this method, the solder joint is broken up into crack zones with relatively constant strain in each zone (see Figure 5-6). The analytical technique described herein is similar to the Solomon method except that only the third zone is considered in the life analysis. Better correlation has been obtained with test data [38] by using only the third zone. Values of the strain factor for the third zone are determined by linear finite-element analysis (see Table 5-2). It is important to note that these values are based upon a typical (50-mil pitch) solder joint as analyzed in Solomon, et. al. [37]. If a different geometry is used, the values would need to be recalculated using finite-element analysis. Since a ball-grid array does not have a fillet, the strain factor is 1.00 (see Table 5-2).

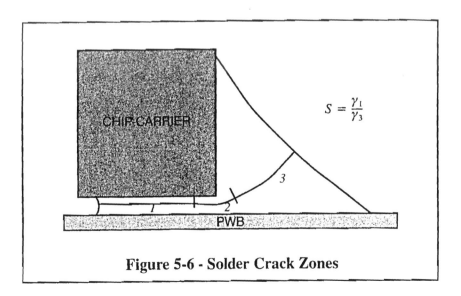

$$S = \frac{\gamma_1}{\gamma_3}$$

**Figure 5-6 - Solder Crack Zones**

| Table 5-2 - Solder Strain Factors (S) | | |
|---|---|---|
| **LCC Joint Configuration** | **Solder Height (h, in)** | **Strain Factor (S)** |
| Minimal 50-mil Pitch [37] | 0.003 | 5.00 |
| Nominal 50-mil Pitch [37] | 0.003 | 6.56 |
| Bulbous 50-mil Pitch [37] | 0.003 | 10.00 |
| Ball–grid array | - | 1.00 |

### Leaded Chip-Carriers Vibration Stress

The vibration stress in a leaded chip-carrier solder joint is given as follows:

$$\tau = \sqrt{\tau_{ip}^2 + \sigma_z^2} \tag{5-21}$$

$$\tau_{ip} = \frac{F_{ip}}{A} \tag{5-22}$$

$$F_{ip} = \max\left(\sqrt{F_{ax}^2 + F_{tx}^2}, \sqrt{F_{ty}^2 + F_{ay}^2}\right) \tag{5-23}$$

$$F_{ax} = K_a\delta_x + K_{al}\delta_y + K_{av}\delta_z \tag{5-24}$$

$$F_{ay} = K_{al}\delta_x + K_a\delta_y + K_{av}\delta_z \tag{5-25}$$

$$F_{tx} = K_{al}\delta_x + K_l\delta_y + K_v\delta_z \tag{5-26}$$

$$F_{ty} = K_l\delta_x + K_{al}\delta_y + K_v\delta_z \tag{5-27}$$

$$\delta_{x,y} = \epsilon_{x,y}\frac{L_{x,y}}{2} \tag{5-28}$$

$$\sigma_z = \max\left(\frac{F_{vx}}{A}, \frac{F_{vy}}{A}\right) \tag{5-29}$$

$$F_{vx} = K_v\,\delta_z + K_{va}\delta_x + K_{vl}\delta_y \tag{5-30}$$

$$F_{vy} = K_v\,\delta_z + K_{vl}\delta_x + K_{va}\delta_y \tag{5-31}$$

$$\delta_z = \max\left(\frac{L_x^2}{16}\mathcal{C}_x, \frac{L_y^2}{16}\mathcal{C}_y\right) \tag{5-20}$$

$\tau$ = *Solder Shear Stress*

$\tau_{ip}$ = *In-Plane Shear Stress*

$\sigma_z$ = *Out-of-Plane Stress*

$F_{ip}$ = *In-Plane Lead Force*

$A$ = *Effective Minimum Load-Bearing Solder Joint Area (see text)*

$F_{ax}, F_{ty}$ = *Axial, Transverse Lead Forces for Lead Aligned with X-axis*

$F_{ay}, F_{tx}$ = *Axial, Transverse Lead Forces for Lead Aligned with Y-axis*

$K_{a,t,v}$ = *Lead Stiffness in Axial, Transverse, Vertical Direction (see text)*

$K_{at}, K_{vt}, K_{va}$ = *Lead Stiffness Cross Terms (see text)*

$\delta_{x,y}$ = *Deflection in x,y Direction*

$\epsilon_{x,y}$ = *PWB Surface Vibration Strain in x,y Direction*

$L_{x,y}$ = *Length of Package Footprint in x,y Direction*

$F_{vx,vy}$ = *Out-of-Plane (vertical) Lead Force for Lead Aligned with X,Y-axis*

$\delta_z$ = *Maximum Out-of-Plane Deflection*

$\mathcal{C}_{x,y}$ = *Curvature in x,y Direction (see 4.5.3)*

It is important to note that the lead forces are different for leads whose axial direction aligns with the x and y axes. The above equations consider the maximum (worst-case) lead force, but if a component has leads on only two sides, only the terms appropriate to that condition need be considered.

<u>Solder Joint Area (A)</u>

The solder joint area is taken as 2/3 of the wetted solder joint area projected to the solder pad. [36]

<u>Lead Stiffnesses ($K_{a,t,v}$, $K_{at,vt,va}$)</u>

The lead stiffnesses represent the forces developed in axial, transverse, and vertical directions per unit deflection in axial, transverse, and vertical directions. Stiffness terms with a single subscript ($K_a$, $K_t$, $K_v$) represent the stiffnesses in the direction of the applied deflection. Stiffness terms with two subscripts represent the stiffness interactions between the directions. Calculation of the stiffness should represent the actual configuration as closely as possible, including the constraint of the lead on the pad due to the solder. Stiffness for various geometries may be determined by classical or finite-element analysis (see Figure 5-7 and 4.7.1).

**Vibration Life Relationship**

The specific life relationship based upon vibration stress is given by adapting equation 5-1 for shear stress and including random vibration as follows:

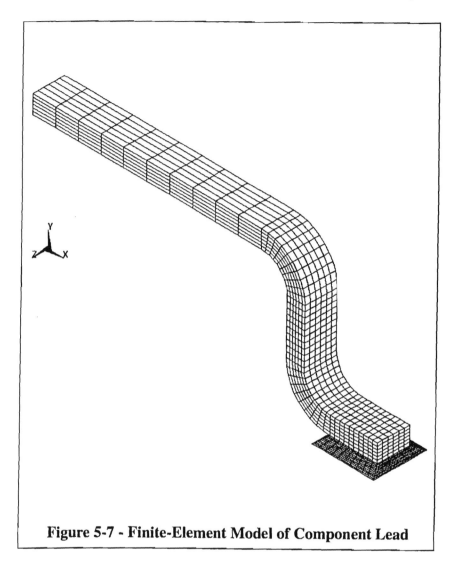

**Figure 5-7 - Finite-Element Model of Component Lead**

$$n_f = \left(\frac{K'}{\tau}\right)^{\lambda} \text{ for Sinusoidal Vibration (use peak stress)}$$ (5-32)

$$n_f = \frac{(K')^{\lambda}}{(0.683 + 0.271(2)^{\lambda} + 0.0456(3)^{\lambda}) \; \tau^{\lambda}} \text{ for Random (RMS)}$$

$$N_F = K_R \, n_f$$ (5-33)

$$T_F = \frac{N_F}{f_n}$$ (5-34)

$n_f$ = Uncorrected Cycles-to-Fail
$K'$ = Solder Vibration Life Multiplier (see Table 5-3)
$\tau$ = Solder Shear Stress
$\lambda$ = Solder Vibration-Life Exponent (see Table 5-3)
$N_F$ = Corrected Cycles-to-Fail
$K_R$ = Reliability Correction Factor (see equation 5-36)
$T_F$ = Time to Failure
$f_n$ = Dominant Resonant Frequency

## Table 5-3 - Solder Vibration Fatigue Parameters [37], [43]

| Solder Alloy | Shear Modulus (MPa) | Fatigue Multiplier (K', MPa) | Fatigue Exponent ($\lambda$) |
|---|---|---|---|
| Tin/Lead (63/37) | 9,192 | 233 | 4.00 |

**Reliability Correction**

The analytical solder-life models predict the mean cycles to failure, which indicates that approximately 50% of the solder joints will fail. In most applications we are particularly interested in where failures start. Literature has indicated that solder joints follow a Weibull failure distribution.[33] The Weibull distribution can be used to develop a reliability correction factor based upon the desired reliability level as follows: [40]

$$F = 1 - e^{-\left(\frac{x}{a'}\right)^{\beta}}$$ (5-35)

Solving for $x$:

$$K_R = x = \alpha' \, (-\ln(1 - F))^{\frac{1}{\beta}} \qquad (5\text{-}36)$$

$F = Cumulative\ Failure\ Probability$

$x = Weibull\ Distributed\ Variable$

$\alpha' = Weibull\ Scale\ Parameter\ (see\ Table\ 5\text{-}4)$

$\beta = Weibull\ Shape\ Parameter\ (see\ Table\ 5\text{-}4)$

$K_R = Reliability\ Correction\ Factor$

The Weibull parameters depend upon the type of stress (thermal cycle or vibration), the type of component (leaded or leadless), and the number of terminals closest to the package corners (number of "corner" terminals). A typical leadless or leaded chip-carrier has leads on four sides and thus has 8 leads exposed to maximum stress. A small-outline integrated circuit (SOIC) has leads on two sides and has 4 leads exposed to maximum stress, and a chip component has only two terminals so it only has 2 terminals at maximum stress.

### Table 5-4 - Parameters for Reliability Correction Factor ($K_R$) [36]

| Data Description | Number of "Corner" Terminals | Weibull Parameters | |
|---|---|---|---|
| | | Alpha ($\alpha'$) | Beta ($\beta$) |
| IPC Leadless Thermal Cycling Model | 8 | 1.096 | 4.000 |
| | 4 | 1.303 | 4.000 |
| | 2 | 1.550 | 4.000 |
| IPC Leaded Thermal Cycling Model | 8 | 1.201 | 2.000 |
| | 4 | 1.698 | 2.000 |
| | 2 | 2.402 | 2.000 |
| IPC Vibration Model | 8 | 1.130 | 3.000 |
| | 4 | 1.424 | 3.000 |
| | 2 | 1.794 | 3.000 |

**Underlying Assumptions**

- Life of solder under vibration is governed by high-cycle SN curve relationship
- Creep of solder under vibration is negligible

## 5.2.3    Thermal Cycling

**Low-Cycle Fatigue Background**

Traditionally, low-cycle fatigue is thought to be governed by the Manson-Coffin relationship, which relates the cycles to failure to the applied plastic strain. The Manson-Coffin approach has been used [33] in the prediction of the fatigue life of solder joints, although the Manson-Coffin approach is a subset of a more generalized low-cycle fatigue model based upon strain energy. Studies have indicated [34] that the traditional Manson-Coffin approach tended to overestimate the solder fatigue life at low strain ranges. Further investigation [35] indicated that energy to fail for eutectic solders near 30°C increased as a function of cycles applied below 2,500 cycles but was constant above 2,500 cycles (see Figure 5-8). This energy relationship is consistent with the deviation of the Manson-Coffin approach at low strains.

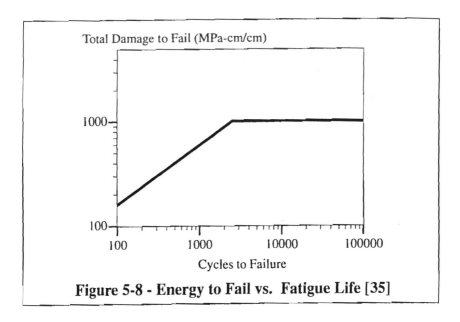

**Figure 5-8 - Energy to Fail vs. Fatigue Life [35]**

**Approach**

<u>Leadless Chip-Carriers and Ball-Grid Arrays</u>

1. Calculate the shear strain in the solder based upon the solder joint geometry, the thermally-induced strain on the surface of the substrate, and a factor for the solder to package mismatch

2. Determine the cycles to fail in the lowest strained zone by applying low-cycle fatigue analysis

<u>Leaded Chip-Carriers</u>

1. Determine the elastic energy in the lead and solder joint assembly

2. Compare the elastic energy with the energy to fail for the solder joint to determine cycles to fail

<u>Both Leaded and Leadless Chip-Carriers</u>

3. Correct the cycles to fail by a factor to represent the change in slope of the solder fatigue curve over 2,500 cycles

4. Correct the cycles to fail by a factor to represent solder reliability

**LCC and Ball-Grid Array Thermal Cycling Life**

The thermal cycling life of leadless chip-carriers (LCCs) and ball-grid array solder joints using near euthenic tin-lead solder is given as follows [33]:

$$n_f = \frac{1}{2}\left(\frac{\varDelta\gamma}{2\epsilon_f'}\right)^{\frac{1}{c}} \tag{5-37}$$

$$c = -0.442 - 6 \times 10^{-4} \, \overline{T}_{SJ} + 0.0174\ln\left(1 + \frac{360}{t_D}\right) \tag{5-38}$$

$$\varDelta\gamma = \frac{\sqrt{(L_x\varDelta a_x \, \varDelta T + a_C \, L_x\varDelta T_C)^2 + (L_y\varDelta a_y \, \varDelta T + a_C \, L_y\varDelta T_C)^2}}{2h \, S} + \varDelta a_{SM} \, \varDelta T \tag{5-39}$$

$$\varDelta a_{x,y} = |a_{x,y} - a_c| \tag{5-40}$$

$$N_F = K_R \, K_L \, n_f \tag{5-41}$$

$n_f$ = Uncorrected Cycles to Fail

$\Delta\gamma$ = Shear Strain in Last Zone to Crack

$2\epsilon_f'$ = Fatigue Ductility Coefficient $\approx$ 0.65

$c$ = Fatigue Ductility Exponent

$\overline{T}_{SJ}$ = Average Solder Temperature ($^\circ C$)

$t_D$ = Half-Cycle Dwell Time (min)

$L_{x,y}$ = Length of Package in x,y Direction

$h$ = Solder Joint Height

$S$ = Strain Factor for Last Zone to Crack  (see Table 5-2)

$\Delta\alpha_{x,y}$ = CTE Mismatch Between Component and Substrate in x,y Direction

$\alpha_{x,y}$ = Substrate (PWB) CTE in x,y Direction

$\alpha_C$ = Component CTE (typically 6ppm/$^\circ C$ for ceramic)

$\Delta T_C$ = Temperature Rise of Component Due to Power

$\Delta\alpha_{SM}$ = Mismatch Between Solder and Component

$\Delta T$ = Thermal Cycling Range

$N_F$ = Corrected Cycles to Fail

$K_L$ = Life Correction Factor (see equation 5-51)

$K_R$ = Reliability Correction Factor  (see equation 5-36)

## Solder-to-Package Mismatch

The mismatch between the package and solder is incorporated by adding a strain based upon this mismatch to the strain in the third zone. This strain depends on the solder and package expansion, and the geometry of the solder joint. This mismatch is obtained empirically or by finite-element analysis.

## Leaded Chip-Carriers Thermal Cycling Analysis

Thermal cycling analysis of leaded chip-carrier solder joints assumes that solder creep causes the elastic energy in the lead and solder joint assembly to be ultimately dissipated in the solder joint [36]. If large temperature ranges or high lead stiffness are present, the solder yields before the full elastic force can be developed. Therefore proper accounting must be made for solder behavior. To consider the yielding of the solder, both the elastic and plastic energy in the solder joint are calculated. This energy is then compared to the energy to fail in the solder joint. The energy to fail in the solder joint is determined by multiplying the maximum energy density in the solder by the effective solder volume:

$$n_f = \frac{1}{2}\left(\frac{U_f}{U_t}\right)^{-\frac{1}{c}} \tag{5-42}$$

$$U_t = \frac{1}{2}K_d \, \delta^2 \quad (for \; \delta \leq \delta_c \; , \; elastic)$$
$$U_t = \frac{1}{2}K_d \, \delta_c^{\; 2} + K_d \, \delta_c \, (\delta - \delta_c) \quad (for \; \delta > \delta_c \; , \; elastic{-}plastic) \tag{5-43}$$

$$K_d = \max\left(\frac{K_a\delta_x^{\;2} + 2K_{at}\delta_x\delta_y + K_t\delta_y^{\;2}}{\delta_x^{\;2} + \delta_y^{\;2}} \; , \; \frac{K_t\delta_x^{\;2} + 2K_{at}\delta_x\delta_y + K_a\delta_y^{\;2}}{\delta_x^{\;2} + \delta_y^{\;2}}\right) \tag{5-44}$$

$$U_f = (2\epsilon_f' \; \tau_c) \, (A \; h) \tag{5-45}$$

$$\delta_c = \frac{\tau_c \, A}{K_d} \tag{5-46}$$

$$\delta = \sqrt{\delta_x^{\;2} + \delta_y^{\;2}} \tag{5-47}$$

$$\delta_{x,y} = \frac{L_{x,y}\Delta a_{x,y} \, \Delta T + a_C \, L_{x,y}\Delta T_C}{2} \tag{5-48}$$

$$\Delta a_{x,y} = |a_{x,y} - a_c| \tag{5-40}$$

$$c = -0.442 - 6 \times 10^{-4} \, \overline{T}_{SJ} + 0.0174 \ln\left(1 + \frac{360}{t_D}\right) \tag{5-38}$$

$$N_F = K_R \, K_L \, n_f \tag{5-41}$$

$n_f = $ Uncorrected Cycles to Fail
$U_f = $ Total Energy to Fail
$U_t = $ Total (elastic + plastic) Energy in Solder Joint
$c = $ Fatigue Ductility Exponent
$K_d = $ Diagonal Flexural Stiffness of Lead + Solder Joint
$K_{at} = $ Lead Stiffness in Axial, Transverse Direction (see text)
$K_{at} = $ Lead Stiffness Cross Term (see text)
$\delta = $ Total Thermal Deflection
$\delta_{x,y} = $ Total Thermal Deflection in x, y Direction
$\delta_c = $ Critical (yield) Deflection
$2\epsilon_f' = $ Fatigue Ductility Coefficient $\approx 0.65$
$\tau_c = $ Critical (yield) Shear Stress (41.4 MPa, see text)
$A = $ Effective Minimum Load-Bearing Solder Joint Area (see text)
$h = $ Solder Joint Height (see text)

$L_{x,y}$ = *Package Length in x,y Direction*

$\Delta a_{x,y}$ = *CTE Mismatch Between Component and Substrate in x,y Direction*

$\Delta T$ = *Thermal Cycling Range*

$a_{x,y}$ = *Substrate (PWB) CTE in x,y Direction*

$a_c$ = *Component CTE (typically 6ppm/°C for ceramic)*

$\Delta T_c$ = *Temperature Rise of Component due to Power*

$\bar{T}_{SJ}$ = *Average Solder Temperature (°C)*

$t_D$ = *Half-Cycle Dwell Time (min)*

$N_F$ = *Corrected Cycles to Fail*

$K_R$ = *Reliability Correction Factor (see equation 5-36)*

$K_L$ = *Life Correction Factor (see equation 5-51)*

As in the vibration model, the lead forces are different for leads whose axial direction aligns with the x and y axes. If a component has leads on two sides, only the terms appropriate to that condition need be considered.

## Critical Shear Stress ($\tau_c$)

The critical yield shear stress factor ($\tau_c$) is used to determine the transition between elastic and plastic deflection. The shear strength of 63/37 solder at room temperature was determined by Wild [39] to be 41.4 MPa (6,000 psi). This 41.4 MPa value produces good agreement with test data [44] if constrained (soldered) lead stiffness is used.

## Solder Joint Area (A) and Height (h)

The solder joint area is taken as 2/3 of the wetted solder joint area projected to the solder pad. The solder joint height is taken as one half the solder paste stencil depth. [36]

## Lead Diagonal Stiffness ($K_d$)

Calculation of the diagonal stiffness should represent the actual configuration as closely as possible. Stiffness may be determined by classical or finite-element analysis.

## Lead Stiffnesses ($K_{a,t}$, $K_{at}$)

The lead stiffnesses represent the forces developed in axial and transverse directions per unit deflection. Stiffness terms with a single subscript ($K_a$, $K_t$) represent the stiffnesses in the direction of the applied deflection. The $K_{at}$ term represents the stiffness interaction between the axial and transverse direction but it is zero for most lead configurations. Calculation of the stiffness should represent the actual configuration as closely as possible, including the constraint of the lead on the pad due to the solder. Stiffness for various geometries may be determined by classical or finite-element analysis (see Figure 5-7 and section 4.7.1).

**Life Correction**

Thermal cycling fatigue of solder is mainly due to low-cycle fatigue as the solder is exposed to thermally-induced strain. Traditionally this fatigue is thought to be governed by the Manson-Coffin low-cycle fatigue relationship, which relates the cycles to failure to the applied plastic strain. The Manson-Coffin approach is a subset of a more generalized low-cycle fatigue model based upon strain energy· [33]

Studies have indicated [35] that the traditional Manson-Coffin approach tended to overestimate the solder fatigue life at low strain ranges. Further investigation indicated that damage (energy) to fail for eutectic solders near 30°C increased as a function of cycles applied below 2,500 cycles but was constant above 2,500 cycles (see Figure 5-8). This deviation is used to develop a life correction factor by using the damage to fail relationship as follows:

$$D_f = D_c \; n_f = K \; (n_f)^{\frac{1}{3}} \quad (\text{for } n_f < 2500) \tag{5-49}$$

$$D_f = D_c \; n_{f \geq 2500} = K \; (2500)^{\frac{1}{3}} \quad (\text{for } n_f \geq 2500) \tag{5-50}$$

Dividing the above equations:

$$K_L = \frac{n_{f \geq 2500}}{n_f} = \left(\frac{2500}{n_f}\right)^{\frac{1}{3}} \; (\text{for } n_f \geq 2500)$$

$$K_L = 1 \; (\text{for } n_f < 2500) \tag{5-51}$$

$D_f = Total \; Damage \; (energy) \; to \; Failure$

$D_c = Damage \; (energy) \; per \; Cycle$

$n_f = Uncorrected \; Cycles \; to \; Fail$

$n_{f \geq 2500} = Life \; Corrected \; Total \; Cycles \; to \; Fail$

$K = Proportionality \; Constant$

$K_L = Life \; Correction \; Factor$

**Determination of Module CTE**

In order to determine the differential expansion between the component and the module surface, the coefficients of thermal expansion (CTEs) of the part and the module surface must be determined. The expansion of ceramics and lead materials are frequently published, and may be used to determine the component expansion. Expansion of the module surface may be determined by measurement [41],[42] using the differential thermally-induced strain between a reference material and the material under test. Module expansion may also be determined by analysis, since the silicone bond layer is very compliant; a finite-element model is usually required (see 4.7.2) if such a bonding method is used.

**Underlying Assumptions**

- Creep performance of solder under low-cycle fatigue is governed by Engelmaier [33] solder fatigue relationships

- Solder is near-eutectic tin-lead

- Since a typical surface-mount module utilizes balanced construction, the out-of-plane deflection (warpage) due to thermal cycling is negligible

- Solder creep causes the elastic energy in a leaded component to be ultimately dissipated in the lead

# 5.3   OTHER LIFE ANALYSIS

Although much of the concern over life analysis in electronic packaging configurations applies to solder joints, life analysis concerns also apply to mechanical hardware, component leads, plated-through holes (PTHs), etc.

## 5.3.1   Mechanical Hardware Life

Vibration and shock loads produce stresses in chassis assemblies and other components which can lead to fatigue failure if the stress levels are sufficiently high. The life of the mechanical hardware in an electronic assembly is determined by using the stresses from finite-element analysis (see Table 4-5) in the generic fatigue equations (5-1 and 5-6).

Low-cycle fatigue is usually not a concern under dynamic loading because yielding does not typically occur in the material. In most thermal cycling situations and where unusually large vibration and/or shock amplitudes are present a low-cycle fatigue analysis is typically required. Because yielding is present, low cycle fatigue analysis requires a non-linear finite-element model.

## 5.3.2   PTH Life Analysis

Typically, a printed wiring board (PWB) has a relatively high coefficient of thermal expansion (CTE) in the out-of-plane direction. Part of this high expansion is due to the expansion rate of the resin used in the PWB and part is due to the Poisson's ratio effect caused by the in-plane reinforcing fibers. This relatively high expansion can cause low-cycle fatigue of the plated-through hole (PTH) barrels. One method of alleviating this low-cycle fatigue is the use of another material (such as nickel) as an overplate to reinforce the copper plating. Analysis of this overplated configuration is somewhat more complicated than a model considering only copper plating [46]. Stresses in the copper and nickel depend upon which of the materials are yielded. If no yielding is present the following model is used:

$$\sigma_{Ni} = \frac{A_{Cu}E_{Cu}(\alpha_{Cu} - \alpha_{Ni}) + A_BE_B(\alpha_B - \alpha_{Ni})}{A_{Ni}E_{Ni} + A_{Cu}E_{Cu} + A_BE_B} E_{Ni} \Delta T \qquad (5\text{-}52)$$

$$\sigma_{Cu} = \frac{A_{Ni}E_{Ni}(\alpha_{Ni} - \alpha_{Cu}) + A_BE_B(\alpha_B - \alpha_{Cu})}{A_{Ni}E_{Ni} + A_{Cu}E_{Cu} + A_BE_B} E_{Cu} \Delta T \qquad (5\text{-}53)$$

$$A_{Cu} = \frac{\pi}{4}\left(d^2 - (d - 2t_{Cu})^2\right) \qquad (5\text{-}54)$$

$$A_{Ni} = \frac{\pi}{4}\left((d - 2t_{Cu})^2 - (d - 2t_{Cu} - 2t_{Ni})^2\right) \qquad (5\text{-}55)$$

$$A_B = \frac{\pi}{4}\left((d + C\ h)^2 - d^2\right) \qquad (5\text{-}56)$$

$$\Delta\epsilon_{Ni} = \frac{\sigma_{Ni}}{E_{Ni}} \qquad (5\text{-}57)$$

$$\Delta\epsilon_{Cu} = \frac{\sigma_{Cu}}{E_{Cu}} \qquad (5\text{-}58)$$

$\sigma_{Ni}, \sigma_{Cu}$ = Stress in Nickel, Copper

$E_{Cu}, E_{Ni}, E_B$ = Modulus of Elasticity of Copper, Nickel, PWB

$A_{Cu}, A_{Ni}$ = Cross-Sectional Area of Copper, Nickel

$\alpha_{Cu}, \alpha_{Ni}, \alpha_B$ = CTE of Copper, Nickel, PWB

$A_B$ = Effective Cross-Sectional Area of PWB

$d$ = Drilled PTH Diameter

$t_{Cu}, t_{Ni}$ = Thickness of Copper, Nickel Plating

$h$ = PWB Thickness

$C$ = Empirical Proprtionality Constant (dimensionless) $\approx$ 1

$\Delta\epsilon_{Ni}, \Delta\epsilon_{Cu}$ = Total Strain in Nickel, Copper

If there is yielding in both the copper and the nickel with a bilinear stress-strain model (see Figure 5-9), the stresses are calculated as follows:

$$\sigma_{Ni} = \frac{A_{Cu}E'_{Cu}(\alpha_{Cu} - \alpha_{Ni}) + A_B E_B(\alpha_B - \alpha_{Ni})}{A_{Ni}E'_{Ni} + A_{Cu}E'_{Cu} + A_B E_B} \; E'_{Ni} \, \Delta T$$
$$+ \frac{A_{Cu}E'_{Cu}(D_{Cu} - D_{Ni}) - A_B E_B D_{Ni}}{A_{Ni}E'_{Ni} + A_{Cu}E'_{Cu} + A_B E_B} \; E'_{Ni} \tag{5-59}$$

$$\sigma_{Cu} = \frac{A_{Ni}E'_{Ni}(\alpha_{Ni} - \alpha_{Cu}) + A_B E_B(\alpha_B - \alpha_{Cu})}{A_{Ni}E'_{Ni} + A_{Cu}E'_{Cu} + A_B E_B} \; E'_{Cu} \, \Delta T$$
$$+ \frac{A_{Ni}E'_{Ni}(D_{Ni} - D_{Cu}) - A_B E_B D_{Cu}}{A_{Ni}E'_{Ni} + A_{Cu}E'_{Cu} + A_B E_B} \; E'_{Cu} \tag{5-60}$$

$$D_{Ni} = S_{y_{Ni}} \frac{E'_{Ni} - E_{Ni}}{E'_{Ni} \, E_{Ni}} \tag{5-61}$$

$$D_{Cu} = S_{y_{Cu}} \frac{E'_{Cu} - E_{Cu}}{E'_{Cu} \, E_{Cu}} \tag{5-62}$$

$$\Delta \epsilon_{Ni} = \frac{S_{y_{Ni}}}{E_{Ni}} + \frac{\sigma_{Ni} - S_{y_{Ni}}}{E'_{Ni}} \tag{5-63}$$

$$\Delta \epsilon_{Cu} = \frac{S_{y_{Cu}}}{E_{Cu}} + \frac{\sigma_{Cu} - S_{y_{Cu}}}{E'_{Cu}} \tag{5-64}$$

$E'_{Cu}, E'_{Ni} = $ *Modulus of Plasticity of Copper, Nickel*
$D_{Cu}, D_{Ni} = $ *Plastic Strain Intercept of Copper, Nickel*
$S_{y_{Cu}}, S_{y_{Ni}} = $ *Yield Stress of Copper, Nickel*

If only the nickel is yielded:

$$\sigma_{Ni} = \frac{A_{Cu}E_{Cu}(\alpha_{Cu} - \alpha_{Ni}) + A_B E_B(\alpha_B - \alpha_{Ni})}{A_{Ni}E'_{Ni} + A_{Cu}E_{Cu} + A_B E_B} \; E'_{Ni} \, \Delta T$$
$$- \frac{A_{Cu}E_{Cu} + A_B E_B}{A_{Ni}E'_{Ni} + A_{Cu}E_{Cu} + A_B E_B} \; E'_{Ni} \, D_{Ni} \tag{5-65}$$

$$\sigma_{Cu} = \frac{A_{Ni}E'_{Ni}(\alpha_{Ni} - \alpha_{Cu}) + A_B E_B(\alpha_B - \alpha_{Cu})}{A_{Ni}E'_{Ni} + A_{Cu}E_{Cu} + A_B E_B} \; E_{Cu} \, \Delta T$$
$$+ \frac{A_{Ni}E'_{Ni}}{A_{Ni}E'_{Ni} + A_{Cu}E_{Cu} + A_B E_B} \; E_{Cu} \, D_{Ni} \tag{5-66}$$

$$\Delta\epsilon_{Cu} = \frac{\sigma_{Cu}}{E_{Cu}} \tag{5-58}$$

$$\Delta\epsilon_{Ni} = \frac{S_{y_{Ni}}}{E_{Ni}} + \frac{\sigma_{Ni} - S_{y_{Ni}}}{E'_{Ni}} \tag{5-63}$$

If only the copper is yielded:

$$\sigma_{Ni} = \frac{A_{Cu}E'_{Cu}(\alpha_{Cu} - \alpha_{Ni}) + A_B E_B(\alpha_B - \alpha_{Ni})}{A_{Cu}E'_{Cu} + A_{Ni}E_{Ni} + A_B E_B} E_{Ni} \Delta T$$

$$+ \frac{A_{Cu}E'_{Cu}}{A_{Cu}E'_{Cu} + A_{Ni}E_{Ni} + A_B E_B} E_{Ni} D_{Cu} \tag{5-67}$$

$$\sigma_{Cu} = \frac{A_{Ni}E_{Ni}(\alpha_{Ni} - \alpha_{Cu}) + A_B E_B(\alpha_B - \alpha_{Cu})}{A_{Cu}E'_{Cu} + A_{Ni}E_{Ni} + A_B E_B} E'_{Cu} \Delta T$$

$$- \frac{A_{Ni}E_{Ni} + A_B E_B}{A_{Cu}E'_{Cu} + A_{Ni}E_{Ni} + A_B E_B} E'_{Cu} D_{Cu} \tag{5-68}$$

$$\Delta\epsilon_{Ni} = \frac{\sigma_{Ni}}{E_{Ni}} \tag{5-57}$$

$$\Delta\epsilon_{Cu} = \frac{S_{y_{Cu}}}{E_{Cu}} + \frac{\sigma_{Cu} - S_{y_{Cu}}}{E'_{Cu}} \tag{5-64}$$

The fatigue life of metals is described by Engelmaier [46] as follows:

$$n_f^{-0.6}D_f^{0.75} + 0.9\frac{S_u}{E}\left(\frac{e^{(D_f)}}{0.36}\right)^{-0.1785\log\frac{10^5}{N_f}} = \Delta\epsilon_{max} \tag{5-69}$$

$$\Delta\epsilon_{max} = K_\epsilon \Delta\epsilon \tag{5-70}$$

$n_f$ = Mean Cycles to Failure

$D_f$ = Ductility (plastic strain at failure)

$S_u$ = Tensile (ultimate) Strength

$E$ = Modulus of Elasticity

$\Delta\epsilon$ = Total Cyclic Strain Range (for both nickel and copper)

$\Delta\epsilon_{max}$ = Effective Maximum Strain Range

$K_\epsilon$ = Strain Concentration Coefficient

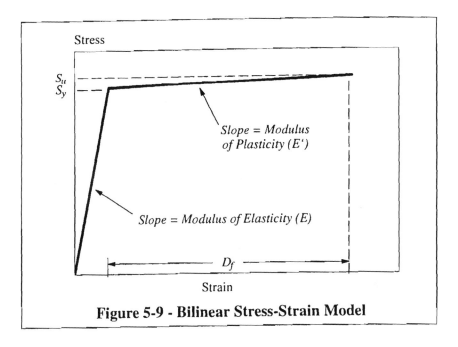

**Figure 5-9 - Bilinear Stress-Strain Model**

It is important to note that the same equation (5-69) is used for the life of both copper and nickel (with appropriate material properties). It is also significant that the equation is non-linear and must be solved numerically for cycles to failure. The strain concentration coefficient ($K_\varepsilon$) results from stress concentrations due to uneven PTH geometry, the influence of internal planes, etc. If zero is used for the area of the nickel ($A_{Ni}$), the equations for stress in the copper agree with those in Reference [46].

The above equations represent a considerable simplification from actual PTH configurations, but results should be adequate for assessing the relative life between various configurations. If more accurate results are required, a finite-element model may be used to determine the stresses and strains in the PTH barrels. Such a model would typically require a non-linear finite-element approach using the stress-strain relationship shown in Figure 5-9. Although the use of non-linear material properties increases complexity, some simplification may be realized by modeling the PTH with 2-dimensional axisymmetric elements. Once the model is developed, the results may be used to directly determine life using equation 5-69, or the model may be used to determine the empirical constants C (in equation 5-56) and $K_\varepsilon$ (in equation 5-70).

**Underlying Assumptions**

- Strain in barrel of PTH is constant through thickness of PWB
- Shear stresses between layers is neglected

# 5.4   COMBINED THERMAL CYCLING AND VIBRATION

## 5.4.1   Loads Applied Independently

If combined thermal cycling and vibration is applied independently (i.e. at different times), the cumulative damage approach (see 5.1.5) is typically used to determine the life of the item in question. In some cases the maximum allowable cumulative damage is reduced below unity (100%) to around 70% for a factor of safety.

## 5.4.2   Loads Applied Simultaneously

In some cases, the combination of thermal cycling and vibration may lead to failure when the the same thermal cycling and vibration levels applied independently do not cause failure. Typically, this failure mode occurs in component leads. This is due to the reduction in fatigue life of a material due to the presence of a residual tensile stress (see 5.1.6). Analytical prediction of fatigue life under combined thermal/vibration cycles involves the following:

1. Calculate the lead deflection based upon the thermal cycling environment

2. Determine the maximum tensile stress by using the lead thermal deflection components (step 1) in a finite-element model of the lead

3. Determine the lead deflection components for the vibration environment

4. Determine the maximum alternating stress by using the lead vibration deflection components (step 3) in a finite-element model of the lead

5. Calculate the equivalent alternating stress using equation 5-15

6. Calculate the lead life based upon the equivalent alternating stress

## Thermal Cycling Deflection Components

Deflection components for thermal cycling are determined by multiplying the relative PWB expansion by the distance of the lead from the package centroid in each direction as follows:

$$\delta_{x,y} = \frac{L_{x,y}\Delta a_{x,y} \, \Delta T + a_C \, L_{x,y}\Delta T_C}{2} \qquad (5\text{-}48)$$

$$\Delta a_{x,y} = |a_{x,y} - a_c| \qquad (5\text{-}40)$$

$\delta_{x,y}$ = *Deflection in x,y Direction*
$L_{x,y}$ = *Length of Component in x,y Direction*
$a_C$ = *Component CTE (typically 6ppm/°C for ceramic)*
$\Delta T$ = *Thermal Cycling Range*
$\Delta T_C$ = *Temperature Rise of Component due to Power*
$\Delta a_{x,y}$ = *Relative PWB CTE in x,y Direction*
$a_{x,y}$ = *PWB CTE in x,y Direction*

If the total deflection due to thermal cycling (see equation 5-47) exceeds the critical deflection (see equation 5-46), the deflection components from equation 5-48 should be multiplied by a factor of $\delta_c/\delta$ to account for yielding of the solder.

## Thermal Cycling Stress

Lead thermal cycling stress is determined by applying the thermal deflection components to a finite-element (or classical) model of the lead to determine stresses in the lead. It is important that the calculation of the lead stresses be consistent with the calculation of the lead stiffnesses and represent the actual configuration as closely as possible (including the effect of the solder joint).

## Vibration Deflection Components

Lead deflection components for vibration are determined using the same approach to calculate the deflection components for solder-life analysis as follows:

$$\delta_{x,y} = \epsilon_{x,y} \frac{L_{x,y}}{2} \qquad (5\text{-}28)$$

$$\delta_z = \max\left( \frac{L_x^2}{16}\mathcal{C}_x \, , \, \frac{L_y^2}{16}\mathcal{C}_y \right) \qquad (5\text{-}20)$$

$\delta_{x,y}$ = *Deflection in x,y Direction*
$\epsilon_{x,y}$ = *PWB Surface Vibration Strain in x,y Direction*
$L_{x,y}$ = *Length of Package Footprint in x,y Direction*
$\delta_z$ = *Maximum Out-of-Plane Deflection*
$\mathcal{C}_{x,y}$ = *Curvature in x,y Direction*

## Vibration Stress

Lead vibration stress is determined by applying the vibration deflection components to a finite-element (or classical) model of the lead to determine stresses. This calculation is essentially the same as the thermal stress calculation although it is important to include the out-of-plane deflection.

## Equivalent Alternating Stress

The equivalent alternating stress is based upon the modified Soderberg relationship described in section 5.1.6:

$$\sigma_{a_{eq}} = \sigma_a \frac{S_y}{S_y - \sigma_m} \qquad (5\text{-}15)$$

$\sigma_{a_{eq}}$ = *Equivalent Alternating Stress*
$\sigma_a$ = *Alternating (vibration) Stress*
$S_y$ = *Yield Stress*
$\sigma_m$ = *Tensile (thermal) Stress*

Using the maximum thermal cycling stress and the maximum vibration stress may be overly conservative if the locations of the maximum thermal and vibration stresses are not the same. In such cases the thermal and vibration stresses may combined at each location of the model to determine the equivalent stress.

## Lead Life

Lead life for sine vibration is determined by using the equivalent alternating stress from equation 5-15 in equation 5-1. For random vibration, life calculation is more complex since the equivalent alternating stress levels are not 1, 2, and 3 times the RMS value, so equations 5-5 and 5-6 no longer apply. Lead life under combined random and tensile stress is determined by calculating the equivalent alternating stress for 1, 2, and 3 times the RMS alternating stress, as follows:

$$n_f = \frac{(K')^\lambda}{0.6826 \; \sigma_1^\lambda + 0.2718 \; \sigma_2^\lambda + 0.0456 \; \sigma_3^\lambda} \tag{5-71}$$

$$\sigma_N = N \; \sigma_{RMS} \frac{S_y}{S_y - \sigma_m} \quad for \; N = 1, 2, 3 \tag{5-72}$$

$n_f$ = *Mean Cycles to Fail*
$K'$ = *Fatigue Life Multiplier*
$\sigma_N$ = *Equivalent Alternating Stress at N times RMS Stress (N = 1,2,3)*
$\lambda$ = *Fatigue Exponent*
$\sigma_{RMS}$ = *Root Mean Square Stress*
$S_y$ = *Yield Stress*
$\sigma_m$ = *Tensile (mean) Stress*

# 5.5   RELATIVE LIFE ANALYSIS

If test data is available for the high or low-cycle fatigue life of a particular component or assembly and life analysis using the methods described above is too complex, a relative life analysis may be used. In a relative life analysis, a characteristic constant is implied that represents a combination of material properties (including fatigue multiplier), and geometry. This characteristic constant is then used to predict the life at a different level.

## 5.5.1   High-Cycle Relative Fatigue Life

**Identical Spectrum Shape (different level)**

High-cycle relative fatigue life for identical spectrum shapes includes random vibration and acoustic spectra with similar frequency rolloff characteristics, and shock pulses with similar shape and duration. In these situations, the input amplitude can be used to scale the expected life, assuming a linear response of the equipment with amplitude, as follows:

$$n_{f_1} = \left(\frac{K'}{\sigma_1}\right)^\lambda$$
$$n_{f_2} = \left(\frac{K'}{\sigma_2}\right)^\lambda \tag{5-1}$$

Dividing the above equations:

$$\frac{n_{f_1}}{n_{f_2}} = \frac{T_{f_1}}{T_{f_2}} = \left(\frac{\sigma_2}{\sigma_1}\right)^\lambda = \left(\frac{A_2}{A_1}\right)^\lambda \qquad (5\text{-}73)$$

For acoustic input:

$$\left(\frac{A_2}{A_1}\right)^\lambda = 10^{\left(\frac{\lambda}{20}\left(SPL_2 - SPL_1\right)\right)} \qquad (5\text{-}74)$$

$n_{f_{1,2}}$ = *Mean Cycles to Fail at Level 1,2*
$K'$ = *Fatigue Life Multiplier*
$\sigma_{1,2}$ = *Peak Stress at Level 1,2*
$\lambda$ = *Fatigue Exponent*
$T_{f_{1,2}}$ = *Time to Fail at Level 1,2*
$A_{1,2}$ = *Input Amplitude at Level 1,2 (see Table 5-5)*
$SPL_{1,2}$ = *Sound Pressure Level 1,2*

| Table 5-5 - Amplitude Measurements for Various Input Types | |
| --- | --- |
| **Type of Input** | **Amplitude Measure (A)** |
| Random Vibration | $G_{RMS}$ |
| Sinusoidal Vibration | G-peak, peat-to-peak deflection |
| Shock | G-peak |
| Acoustic | RMS pressure |
| Other | Relative deflection |

**Varying Spectrum Shape**

If the random or acoustic spectrum shape, or the sinusoidal vibration curve varies a slightly different approach must be used to determine the relative life. Typically the largest relative deflection occurs at the lowest resonant frequency so relative life for a varying spectrum may be determined by using the input levels at this frequency as follows:

$$\frac{n_{f_1}}{n_{f_2}} = \frac{T_{f_1}}{T_{f_2}} = \left(\frac{A_2}{A_1}\right)^\lambda \tag{5-73}$$

For random vibration:

$$\left(\frac{A_2}{A_1}\right)^\lambda = \left(\frac{P_2}{P_1}\right)^{\frac{\lambda}{2}} \tag{5-75}$$

For acoustic input:

$$\left(\frac{A_2}{A_1}\right)^\lambda = 10^{\left(\frac{\lambda}{20}(PSL_2 - PSL_1)\right)} \tag{5-76}$$

$n_{f_{1,2}}$ = Mean Cycles to Fail at Level 1,2
$\lambda$ = Fatigue Exponent
$T_{f_{1,2}}$ = Time to Fail at Level 1,2
$A_{1,2}$ = Input Amplitude at Resonance for Level 1,2
$P_{1,2}$ = Input Power Spectral Density at Resonance for Level 1,2
$PSL_{1,2}$ = Pressure Spectrum at Resonance for Level 1,2

Relative life for shock pulses with varying shapes or duration is complex and needs to be determined by using equation 5-73 with the measured or analytically determined relative deflection for each input pulse.

## 5.5.2   Low-Cycle Relative Fatigue Life

### Generic Low-Cycle Relative Life

Relative life for low-cycle fatigue is determined by comparing the plastic deformation corresponding to each input level. If the life comparison is from a large cycle to a small cycle, the assumption that the plastic deformation is equal to the total deformation is conservative (since the plastic deformation is a smaller fraction of the total for a small cycle). With this assumption in mind and taking the deformation as proportional to the thermal cycling range, the relative life for low-cycle fatigue is as follows:

$$n_{f_1} = \frac{1}{2}\left(\frac{\Delta\epsilon_{p_1}}{2\epsilon_f'}\right)^{\frac{1}{c}}$$

$$n_{f_2} = \frac{1}{2}\left(\frac{\Delta\epsilon_{p_2}}{2\epsilon_f'}\right)^{\frac{1}{c}} \tag{5-2}$$

Dividing the above equations:

$$\frac{n_{f_1}}{n_{f_2}} = \left(\frac{\Delta \epsilon_{p_1}}{\Delta \epsilon_{p_2}}\right)^{\frac{1}{c}} \approx \left(\frac{\Delta \epsilon_{t_1}}{\Delta \epsilon_{t_2}}\right)^{\frac{1}{c}} = \left(\frac{\Delta T_1}{\Delta T_2}\right)^{\frac{1}{c}} \qquad (5\text{-}77)$$

$n_{f_{1,2}}$ = *Mean Cycles to Fail at Level 1,2*
$\Delta \epsilon_{p_{1,2}}$ = *Plastic Strain at Level 1,2*
$2\epsilon_f'$ = *Fatigue Ductility Coefficient*
$c$ = *Fatigue Ductility Exponent*
$\Delta \epsilon_{t_{1,2}}$ = *Total Strain at Level 1,2*
$\Delta T_{1,2}$ = *Thermal Cycling Range at Level 1,2*

**Solder Relative Life**

Relative life for solder is similar to the generic low-cycle fatigue life comparison described above but with the following exceptions:

- Since the fatigue ductility coefficient (c) described by Engelmaier [33] accounts for the relationship between plastic strain and total strain, it is no longer necessary to assume that plastic deformation equals total deformation

- The life correction factor and temperature dependent ductility coefficient (c) make it necessary for the relative life to be based upon a specific number of cycles instead of a life ratio as shown in equation 5-77

Based upon the above, the first step is to "back-out" the life correction factor from the reference cycles to fail to obtain an uncorrected cycles to fail, as follows:

$$N_F = K_R \, K_L \, n_f \qquad (5\text{-}41)$$

$$K_L = \left(\frac{2500}{n_f}\right)^{\frac{1}{3}} \text{ (for } n_f \geq 2500)$$

$$K_L = 1 \text{ (for } n_f < 2500) \qquad (5\text{-}51)$$

If we consider similar reliability for test and evaluation environments:

$$n_{f_1} = \frac{N_{f_1}}{K_L} \qquad (5\text{-}78)$$

$$K_L = \left(\frac{2500}{N_{f_1}}\right)^{\frac{1}{2}} \text{(for } N_{f_1} \geq 2500) \qquad (5\text{-}79)$$

$$K_L = 1 \text{ (for } N_{f_1} < 2500)$$

$K_R$ = *Reliability Correction Factor*
$K_L$ = *Life Correction Factor*
$N_F$ = *Corrected Cycles to Fail*
$n_{f_1}$ = *Uncorrected Cycles to Fail for Reference Environment*
$N_{f_1}$ = *Cycles to Fail for Reference Environment*

Once the life correction factor is accounted for, the approach for relative life analysis is to use the temperature range to establish solder strain and use the solder strain to establish relative solder life. The relationship between life and solder strain is expressed in simplified form as follows:

$$n_f = \frac{1}{2}\left(\frac{\Delta\gamma}{2\epsilon_f'}\right)^{\frac{1}{c}} \qquad (5\text{-}37)$$

$$c = -0.442 - 6 \times 10^{-4}\,\overline{T}_{SJ} + 0.0174\ln\left(1 + \frac{360}{t_D}\right) \qquad (5\text{-}38)$$

$$\Delta\gamma = \frac{L_D}{h\,S}(\Delta\alpha\,\Delta T + \alpha_C\,\Delta T_C) + \Delta\alpha_{SM}\,\Delta T \qquad (5\text{-}80)$$

$n_f$ = *Uncorrected Cycles to Fail*
$\Delta\gamma$ = *Shear Strain*
$2\epsilon_f'$ = *Fatigue Ductility Coefficient*
$c$ = *Fatigue Ductility Exponent*
$\overline{T}_{SJ}$ = *Average Solder Temperature (°C)*
$t_D$ = *Half-Cycle Dwell Time (min)*
$L_D$ = *Diagonal Distance from Package Center*
$h$ = *Solder Joint Height*
$S$ = *Strain Factor*
$\Delta\alpha$ = *CTE Mismatch between Component and Substrate*

$\alpha_C$ = *Component CTE*
$\Delta T_C$ = *Temperature Rise of Component Due to Power*
$\Delta \alpha_{SM}$ = *Mismatch between Solder and Component*
$\Delta T$ = *Thermal Cycling Range*

The relationship between the solder strain and temperature cycling range depends upon which terms from equation 5-80 are considered. If the local temperature rise of the component due to power is very small, the following approach is used:

$$\Delta \gamma = \frac{L_D}{h\,S}(\Delta \alpha \, \Delta T + \alpha_C \; x \; 0) + \Delta \alpha_{SM} \, \Delta T \qquad (5\text{-}80)$$

Expressing equation 5-80 as a proportionality and including the fatigue ductility coefficient in the proportionality constant:

$$\Delta \gamma \propto \Delta T \qquad (5\text{-}81)$$

$$n_f = \frac{1}{2}\left(\frac{\Delta T}{F}\right)^{\frac{1}{c}} \qquad (5\text{-}82)$$

$$F = \frac{\Delta T_1}{\left(2\,n_{f_1}\right)^{c_1}} \qquad (5\text{-}83)$$

$$n_{f_2} = \frac{1}{2}\left(\frac{\Delta T_2}{F}\right)^{\frac{1}{c_2}} \qquad (5\text{-}84)$$

$F$ = *Proportionality Factor*
$n_{f_{1,2}}$ = *Uncorrected Cycles to Fail for Temperature Range 1,2*
$c_{1,2}$ = *Fatigue Ductility Exponent for Temperature Range 1,2*
$\Delta T_{1,2}$ = *Thermal Cycling Range 1,2*

If the temperature rise of the component is not small enough to neglect, the relative life can be determined if the solder-to-component mismatch is neglected and expansion rates of the applicable components are known:

$$\Delta \gamma = \frac{L_D}{h\,S}(\Delta \alpha \, \Delta T + \alpha_C \, \Delta T_C) + 0 \; x \; \Delta T \qquad (5\text{-}80)$$

Expressing equation 5-80 as a proportionality and including the fatigue ductility coefficient in the proportionality constant:

$$\Delta\gamma \propto \Delta a \ \Delta T + a_c \ \Delta T_c \tag{5-85}$$

$$n_f = \frac{1}{2}\left(\frac{\Delta a \ \Delta T + a_c \ \Delta T_c}{F'}\right)^{\frac{1}{c}} \tag{5-86}$$

$$F' = \frac{\Delta a \ \Delta T_1 + a_c \ \Delta T_{c_1}}{\left(2 \ n_{f_1}\right)^{c_1}} \tag{5-87}$$

$$n_{f_2} = \frac{1}{2}\left(\frac{\Delta a \ \Delta T_2 + a_{c_2} \ \Delta T_c}{F'}\right)^{\frac{1}{c_2}} \tag{5-88}$$

$F'$ = Proportionality Factor

$n_{f_{1,2}}$ = Uncorrected Cycles to Fail for Temperature Range 1,2

$c_{1,2}$ = Fatigue Ductility Exponent for Temperature Range 1,2

$\Delta T_{1,2}$ = Thermal Cycling Range 1,2

$\Delta a$ = CTE Mismatch between Component and Substrate

$a_c$ = Component CTE

$\Delta T_{c_{1,2}}$ = Temperature Rise of Component due to Power for Condition 1,2

Once the uncorrected cycles to fail for the proposed environment is determined, the life correction factor is used to determine the cycles to fail:

$$K_L = \left(\frac{2500}{n_{f_2}}\right)^{\frac{1}{3}} \quad (\text{for } n_{f_2} \geq 2500) \tag{5-51}$$

$$K_L = 1 \quad (\text{for } n_{f_2} < 2500)$$

If we consider similar reliability for test and evaluation environments:

$$N_{f_2} = K_L \ n_{f_2} \tag{5-89}$$

$K_L$ = Life Correction Factor

$n_{f_2}$ = Uncorrected Cycles to Fail for Proposed Environment

$N_{f_2}$ = Cycles to Fail for Proposed Environment

It is important to reiterate that since the life correction factor depends upon an absolute number of cycles and the fatigue ductility exponent varies with temperature, the acceleration factor between the environments is not scaleable. Actual test and proposed environment cycle numbers should be used where possible.

### 5.5.3    Underlying Assumptions

- System response is linear for varying input levels

- Resonant frequency does not vary with varying input levels

- Maximum relative deflection occurs at lowest resonant frequency (varying vibration/acoustic spectrum shape)

- Plastic deformation in generic low-cycle fatigue is equal to total deformation

- Total deformation is proportional to thermal cycling range

- Solder to component mismatch is neglected when temperature rise due to power is included

## 5.6    TYPICAL LIFE CALCULATIONS

### 5.6.1    Leadless Solder Joint Thermal Cycling Life

For the surface-mount technology module described in 4.7.2 (see Figure 4-20), determine the life of the solder joints on a 28-terminal leadless chip-carrier (0.45 in x 0.45 in) mounted in the center of the module when exposed to thermal cycling from 0°C to 70°C. Total cycle duration is 1.5 hours with half-hour dwells at the temperature extremes. Component power dissipation is low so the temperature rise of the component relative to the module surface is essentially zero. Tin-lead eutectic solder joints are typical geometry with a height of 0.003 in under the package (50-mil pitch). The desired cumulative failure probability is 1%.

**Uncorrected Thermal Cycling Life**

Using the leadless chip-carrier thermal cycling relationships:

$L_x = L_y = Length\ of\ Package\ in\ x,y\ Direction = 0.45\ in$

$h = Solder\ Joint\ Height = 0.003\ in$

$2\epsilon_f' = Fatigue\ Ductility\ Coefficient \approx 0.65$

$\Delta T = Thermal\ Cycling\ Range = 70°C - 0°C = 70°C$

$\Delta T_C = Temperature\ Rise\ of\ Component\ due\ to\ Power = 0°C$

$\overline{T}_{SJ} = Average\ Solder\ Temperature = \dfrac{70°C + 0°C}{2} = 35°C$

$t_D = Dwell\ Time = 30\ min$

$\alpha_C = Component\ CTE = 6\ ppm/°C$

From Table 5-2 for a nominal 50-mil solder joint with 0.003-in thickness:

$S$ = *Strain Factor for Last Zone to Crack* = 6.56

Combining solder CTE from Table 4-3 and the component CTE:

$\Delta a_{SM}$ = *Solder-Component Mismatch* = $28.3\frac{ppm}{°C} - 6\frac{ppm}{°C} = 22.3\frac{ppm}{°C}$

From Figures 4-22 and 4-23, the CTEs at the center of the module surface are:

$a_x$ = *CTE in x Direction* = $17.2\frac{ppm}{°C}$

$a_y$ = *CTE in y Direction* = $16.4\frac{ppm}{°C}$

$$\Delta a_x = |a_x - a_c| = \left|17.2\frac{ppm}{°C} - 6\,ppm/°C\right| = 11.2\frac{ppm}{°C}$$

$$\Delta a_y = |a_y - a_c| = \left|16.4\frac{ppm}{°C} - 6\,ppm/°C\right| = 10.4\frac{ppm}{°C} \tag{5-40}$$

$$c = -0.442 - 6 \times 10^{-4}\,\overline{T}_{SJ} + 0.0174\ln\left(1 + \frac{360}{t_D}\right) \tag{5-38}$$

$$c = -0.442 - 6 \times 10^{-4}\, x\, 35°C + 0.0174\ln\left(1 + \frac{360}{30\,min}\right) = -0.418$$

$$\Delta\gamma = \frac{\sqrt{(L_x\Delta a_x\,\Delta T + a_c\,L_x\Delta T_c)^2 + (L_y\Delta a_y\,\Delta T + a_c\,L_y\Delta T_c)^2}}{2h\,S} + \Delta a_{SM}\,\Delta T \tag{5-39}$$

$$\Delta\gamma = \frac{\sqrt{\left(0.45inx11.2\frac{ppm}{°C}\,70°C + 0\right)^2 + \left(0.45inx10.4\frac{ppm}{°C}\,70°C + 0\right)^2}}{2\,x\,0.003in\,x\,6.56} + 22.3\frac{ppm}{°C}70°C$$

$$\Delta\gamma = 13,793\,ppm = 0.0138$$

$$n_f = \frac{1}{2}\left(\frac{\Delta\gamma}{2\epsilon_f'}\right)^{\frac{1}{c}} = \frac{1}{2}\left(\frac{0.0138}{0.65}\right)^{\frac{1}{-0.418}} = 5,029\,cycles \tag{5-37}$$

**Life Correction Factor**

$$n_f = Uncorrected\ Cycles\ to\ Fail = 6,753\ cycles$$

$$K_L = \left(\frac{2500}{n_f}\right)^{\frac{1}{3}}\ (for\ n_f \geq 2500)$$

$$K_L = 1\ (for\ n_f < 2500) \tag{5-51}$$

$$K_L = \left(\frac{2500}{n_f}\right)^{\frac{1}{3}} = \left(\frac{2500}{5029}\right)^{\frac{1}{3}} = \mathbf{0.792}$$

**Reliability Correction Factor**

$$F = Cumulative\ Failure\ Probaility = 1\% = 0.01$$

From Table 5-4 for thermal cycling of a leadless chip-carrier:

$$\alpha' = Weibull\ Scale\ Parameter = 1.096$$
$$\beta = Weibull\ Shape\ Parameter = 4.00$$

$$K_R = \alpha'\ (-\ln(1-F))^{\frac{1}{\beta}} = 1.096\ (-\ln(1-0.01))^{\frac{1}{4.00}} = \mathbf{0.347} \tag{5-36}$$

**Corrected Thermal Cycling Life**

The corrected thermal cycling life for the leadless chip-carrier on the module described above is determined to be 1,683 cycles by applying the life and reliability correction factors to the uncorrected life as follows:

$$n_f = Uncorrected\ Cycles\ to\ Fail = 5,029\ cycles$$
$$K_L = Life\ Correction\ Factor = 0.792$$
$$K_R = Reliability\ Correction\ Factor = 0.347$$

$$N_F = K_R\ K_L\ n_f = 0.347\ x\ 0.792\ x\ 5,029\ cycles$$
$$N_F = \mathbf{1,382\ cycles} \tag{5-41}$$

Since the cycle duration is 1.5 hours, the total life under the thermal cycle described above is 2,073 hours.

## 5.6.2    Leadless Solder Joint Vibration Life

For the surface-mount technology module described in 4.7.3 (see Figure 4-20), determine the life of the solder joints on the 28-terminal leadless chip-carrier described in 5.6.1 when exposed to the vibration environment described in 4.7.3. The desired cumulative failure probability is 1%.

## Leadless Chip-Carrier Vibration Stress

Using the leadless chip-carrier vibration-stress relationship:

$L_x = L_y = $ *Length of Package in x,y Direction* $= 0.45$ *in*
$h = $ *Solder Joint Height* $= 0.003$ *in*

From Table 5-2 for a nominal 50-mil solder joint with 0.003-in thickness:

$S = $ *Strain Factor for Last Zone to Crack* $= 6.56$

From Table 5-3:

$G = $ *Solder Shear Modulus* $= 9,192$ *MPa* $= 1.33 \times 10^6$ *psi*

From Figures 4-26 and 4-27, the strains at the center of the module surface are:

$\epsilon_x = $ *PWB Surface Vibration Strain in x Direction* $= 3.62 \times 10^{-5}$
$\epsilon_y = $ *PWB Surface Vibration Strain in y Direction* $= 1.20 \times 10^{-5}$

From Figures 4-29 and 4-30, the curvatures at the center of the module surface are:

$e_x = $ *Curvature in x Direction* $= 1.01 \times 10^{-3}$ *in*$^{-1}$
$e_y = $ *Curvature in y Direction* $= 3.58 \times 10^{-4}$ *in*$^{-1}$

$$\gamma_x = \epsilon_x \frac{L_x}{2h} = \frac{3.62 \times 10^{-5} \times 0.45 \ in}{2 \times 0.003 \ in} = 2.72 \times 10^{-3}$$

$$\gamma_y = \epsilon_y \frac{L_y}{2h} = \frac{1.20 \times 10^{-5} \times 0.45 \ in}{2 \times 0.003 \ in} = 9.00 \times 10^{-4}$$

(5-18)

$$\delta_z = max\left( \frac{L_x^2}{16} e_x , \ \frac{L_y^2}{16} e_y \right)$$

(5-20)

$$\delta_z = max\left( \frac{(0.45 in)^2}{16} 1.01 \times 10^{-3} in^{-1} , \ \frac{(0.45 in)^2}{16} 3.58 \times 10^{-4} in^{-1} \right) = 1.28 \times 10^{-5} \ in$$

$$\epsilon_z = \frac{\delta_z}{h} = \frac{1.28 \times 10^{-5} \ in}{0.003 \ in} = 4.27 \times 10^{-3}$$

(5-19)

$$\gamma = \sqrt{\gamma_x^2 + \gamma_y^2 + \epsilon_z^2}$$

(5-17)

$$\gamma = \sqrt{(2.72 \times 10^{-3})^2 + (9.00 \times 10^{-4})^2 + (4.27 \times 10^{-3})^2} = 5.14 \times 10^{-3}$$

$$\tau = G \frac{\gamma}{S} = 1.33 \times 10^6 \ psi \ 5.14 \times \frac{10^{-3}}{6.56} = 1,042 \ psi$$

(5-16)

**Uncorrected Vibration Life**

$\tau = Solder\ Shear\ Stress = 1,042\ psi$

From Table 5-3:

$K' = Solder\ Vibration\ Life\ Multiplier = 233\ MPa = 33,793\ psi$
$\lambda = Solder\ Vibration\ Life\ Exponent = 4.00$

$$n_f = \frac{(K')^\lambda}{(0.683 + 0.271(2)^\lambda + 0.0456(3)^\lambda)\ \tau^\lambda}$$

$$n_f = \frac{(33,793\ psi)^4}{(0.683 + 0.271(2)^4 + 0.0456(3)^4)\ (1,042\ psi)^4} = \mathbf{1.27x10^5\ cycles}$$

(5-32)

**Reliability Correction Factor**

$F = Cumulative\ Failure\ Probaility = 1\% = 0.01$

From Table 5-4 for vibration of a leadless chip-carrier:

$a' = Weibull\ Scale\ Parameter = 1.130$
$\beta = Weibull\ Shape\ Parameter = 3.00$

$$K_R = a'\ (-\ln(1-F))^{\frac{1}{\beta}} = 1.130\ (-\ln(1-0.01))^{\frac{1}{3.00}} = \mathbf{0.244}$$

(5-36)

**Corrected Vibration Life**

The corrected vibration life for the leadless chip-carrier on the module described above is determined to be 4.8 minutes by applying the reliability correction factor to the uncorrected life as follows:

$n_f = Uncorrected\ Cycles\ to\ Fail = 1.27x10^5\ cycles$
$K_R = Reliability\ Correction\ Factor = 0.244$
$f_n = Dominant\ Resonant\ Frequency = 107\ Hz$

$$N_F = K_R\ n_f = 0.244\ x\ 1.27x10^5\ cycles = \mathbf{3.10x10^4\ cycles}$$

(5-33)

$$T_F = \frac{N_F}{f_n} = \frac{3.10x10^4\ cycles}{107\ Hz} = \mathbf{289\ sec} = \mathbf{4.8\ min}$$

(5-34)

## 5.6.3    Leaded Solder Joint Thermal Cycling Life

For the surface-mount technology module described in 4.7.2 (see Figure 4-20), determine the life of the solder joints on a 28-terminal leaded chip-carrier (0.45 in x 0.45 in) mounted in the center of the module when exposed to thermal cycling from 0°C to 70°C. Total cycle duration is 1.5 hours with half-hour dwells at the temperature extremes. Component power dissipation is low so the temperature rise of the component relative to the module surface is essentially zero. The lead configuration is as described in 4.7.1 (see Figure 4-15) and it is soldered using an 0.008-in thick solder stencil. The desired cumulative failure probability is 1%.

**Uncorrected Thermal Cycling Life**

Using the leaded chip-carrier thermal cycling relationships:

$L_x = L_y = Length\ of\ Package\ in\ x,y\ Direction = 0.45\ in$

$h = Solder\ Joint\ Height = \frac{1}{2} Stencil\ Thickness = 0.004\ in$

$2\epsilon_f' = Fatigue\ Ductility\ Coefficient \approx 0.65$

$\Delta T = Thermal\ Cycling\ Range = 70°C - 0°C = 70°C$

$\Delta T_C = Temperature\ Rise\ of\ Component\ due\ to\ Power = 0°C$

$\overline{T}_{SJ} = Average\ Solder\ Temperature = \dfrac{70°C + 0°C}{2} = 35°C$

$t_D = Dwell\ Time = 30\ min$

$a_C = Component\ CTE = 6\ ppm/°C$

$\tau_c = Critical\ (yield)\ Shear\ Stress = 41.4\ MPa = 6,000psi$

From 4.7.1 the lead stiffness factors are:

$K_a = Lead\ Stiffness\ in\ Axial\ Direction = 20,868\dfrac{lbf}{in}$

$K_t = Lead\ Stiffness\ in\ Transverse\ Direction = 13,065\dfrac{lbf}{in}$

$K_{at} = Lead\ Stiffness\ Cross\ Term = 0$

From Figures 4-22 and 4-23, the CTEs at the center of the module surface are:

$a_x = CTE\ in\ x\ Direction = 17.2\dfrac{ppm}{°C}$

$a_x = CTE\ in\ y\ Direction = 16.4\dfrac{ppm}{°C}$

From Figure 4-15, using outer lead radius of 0.043 in and width of 0.021 in and considering the solder to wet up to 45° from centerline of lead:

$A = Effective\ Solder\ Joint\ Area = \dfrac{2}{3}x\ 0.021in\ x\ 2\sin(45°)\ x\ 0.043in$

$A = 8.51x10^{-4}\ in^2$

$$\Delta a_x = |a_x - a_c| = \left| 17.2\frac{ppm}{°C} - 6\,ppm/°C \right| = 11.2\frac{ppm}{°C}$$

$$\Delta a_y = |a_y - a_c| = \left| 16.4\frac{ppm}{°C} - 6\,ppm/°C \right| = 10.4\frac{ppm}{°C} \tag{5-40}$$

$$c = -0.442 - 6 \times 10^{-4}\,\overline{T}_{SJ} + 0.0174\ln\left( 1 + \frac{360}{t_D} \right) \tag{5-38}$$

$$c = -0.442 - 6 \times 10^{-4}\,x\,35°C + 0.0174\ln\left( 1 + \frac{360}{30\,min} \right) = -0.418$$

$$\delta_x = \frac{L_x\Delta a_x\,\Delta T + a_c\,L_x\Delta T_C}{2} = \frac{0.45inx11.2\frac{ppm}{°C}\ 70°C + 0}{2} = 1.76x10^{-4}\,in \tag{5-48}$$

$$\delta_y = \frac{L_y\Delta a_y\,\Delta T + a_c\,L_y\Delta T_C}{2} = \frac{0.45inx10.4\frac{ppm}{°C}\ 70°C + 0}{2} = 1.64x10^{-4}\,in$$

$$U_f = (2\epsilon_f'\,\tau_c)\,(A\ h) = (0.65\ x\ 6,000\,psi)\,(8.51x10^{-4}\,in^2\ x\ 0.004\,in)$$

$$U_f = 1.33x10^{-2}\,in\text{-}lbf \tag{5-45}$$

$$K_d = \max\left( \frac{K_a\delta_x^2 + 2K_a\delta_x\delta_y + K_l\delta_y^2}{\delta_x^2 + \delta_y^2}\ ,\ \frac{K_l\delta_x^2 + 2K_a\delta_x\delta_y + K_a\delta_y^2}{\delta_x^2 + \delta_y^2} \right)$$

$$\frac{20,868\frac{lbf}{in}(1.76x10^{-4}\,in)^2 + 0 + 13,065\frac{lbf}{in}(1.64x10^{-4})^2}{(1.76x10^{-4}\,in)^2 + (1.64x10^{-4})^2} = 17,242\frac{lbf}{in}$$

$$\frac{13,065\frac{lbf}{in}(1.76x10^{-4}\,in)^2 + 0 + 20,868\frac{lbf}{in}(1.64x10^{-4})^2}{(1.76x10^{-4}\,in)^2 + (1.64x10^{-4})^2} = 16,691\frac{lbf}{in} \tag{5-44}$$

$$K_d = \max\left( 17,242\frac{lbf}{in}\ ,\ 16,691\frac{lbf}{in} \right) = 17,242\frac{lbf}{in}$$

$$\delta_c = \frac{\tau_c\,A}{K_d} = \frac{6,000psi\ x\ 8.51x10^{-4}\,in^2}{17,242\frac{lbf}{in}} = 2.96x10^{-4}\,in \tag{5-46}$$

$$\delta = \sqrt{\delta_x^2 + \delta_y^2} = \sqrt{(1.76x10^{-4}in)^2 + (1.64x10^{-4}in)^2} = 2.41x10^{-4}in \tag{5-47}$$

$$U_t = \frac{1}{2}K_d\,\delta^2 = \frac{17,242}{2}\frac{lbf}{in}(2.41x10^{-4}in)^2 = 5.01x10^{-4}in\text{-}lbf\ (\delta \le \delta_c) \tag{5-43}$$

$$n_f = \frac{1}{2}\left( \frac{U_f}{U_t} \right)^{-\frac{1}{c}} = \frac{1}{2}\left( \frac{1.33x10^{-2}in\text{-}lbf}{5.01x10^{-4}in\text{-}lbf} \right)^{-\frac{1}{0.418}} = \mathbf{1,275\ cycles} \tag{5-42}$$

## Life Correction Factor

$n_f$ = Uncorrected Cycles to Fail = 1, 275 cycles

$$K_L = \left(\frac{2500}{n_f}\right)^{\frac{1}{3}} \text{(for } n_f \geq 2500)$$

$$K_L = 1 \text{ (for } n_f < 2500)$$  (5-51)

$$K_L = 1$$

## Reliability Correction Factor

$F$ = Cumulative Failure Probaility = 1% = 0.01

From Table 5-4 for thermal cycling of a leaded chip-carrier:

$a'$ = Weibull Scale Parameter = 1.201

$\beta$ = Weibull Shape Parameter = 2.00

$$K_R = a' \ (-\ln(1-F))^{\frac{1}{\beta}} = 1.201 \ (-\ln(1-0.01))^{\frac{1}{2.00}} = \mathbf{0.120}$$  (5-36)

## Corrected Thermal Cycling Life

The corrected thermal cycling life for the leaded chip-carrier on the module described above is determined to be 1,683 cycles by applying the life and reliability correction factors to the uncorrected life as follows:

$n_f$ = Uncorrected Cycles to Fail = 1, 275 cycles

$K_L$ = Life Correction Factor = 1.000

$K_R$ = Reliability Correction Factor = 0.120

$$N_F = K_R \ K_L \ n_f = 0.120 \ x \ 1.000 \ x \ 1, 275 \ cycles$$  (5-41)

$$N_F = \mathbf{153 \ cycles}$$

Since the cycle duration is 1.5 hours, the total life under the thermal cycle described above is 230 hours. This example illustrates the unusual situation where a very stiff lead actually reduces the thermal cycling life because there is a smaller solder joint than is the case with a leadless chip-carrier, and the reliability distribution is less favorable.

### 5.6.4    Leaded Solder Joint Vibration Life

For the surface-mount technology module described in 4.7.3 (see Figure 4-20), determine the life of the solder joints on the 28-terminal leaded chip-carrier described in 5.6.3 when exposed to the vibration environment described in 4.7.3. The desired cumulative failure probability is 1%.

**Leaded Chip-Carriers Vibration Stress**

The vibration stress in a leaded chip-carrier solder joint is given as follows:

$L_x = L_y = $ *Length of Package in x,y Direction* $ = 0.45$ *in*

From Figures 4-26 and 4-27, the strains at the center of the module surface are:

$\epsilon_x = $ *PWB Surface Vibration Strain in x Direction* $ = 3.62x10^{-5}$

$\epsilon_y = $ *PWB Surface Vibration Strain in y Direction* $ = 1.20x10^{-5}$

From Figures 4-29 and 4-30, the curvatures at the center of the module surface are:

$\mathcal{C}_x = $ *Curvature in x Direction* $ = 1.01x10^{-3}$ *in$^{-1}$*

$\mathcal{C}_y = $ *Curvature in y Direction* $ = 3.58x10^{-4}$ *in$^{-1}$*

From 4.7.1 the lead stiffness factors are:

$K_a = $ *Lead Stiffness in Axial Direction* $ = 20,868\dfrac{lbf}{in}$

$K_t = $ *Lead Stiffness in Transverse Direction* $ = 13,065\dfrac{lbf}{in}$

$K_v = $ *Lead Stiffness in Vertical Direction* $ = 57,517\dfrac{lbf}{in}$

$K_{av} = K_{va} = $ *Lead Stiffness Vertical-Axial Cross Stiffness* $ = 24,821\dfrac{lbf}{in}$

$K_{at} = K_{ta} = K_{vt} = K_{tv} = $ *Other Lead Stiffness Cross Terms* $ = 0$

From Figure 4-15, using outer lead radius of 0.043 in and width of 0.021 in and considering the solder to wet up to 45° from centerline of lead:

$A = $ *Effective Solder Joint Area* $ = \dfrac{2}{3}x\ 0.021in\ x\ 2\sin(45°)\ x\ 0.043in$

$A = 8.51x10^{-4}$ *in$^2$*

$$\delta_x = \epsilon_x\ \frac{L_x}{2} = 3.62x10^{-5}\ \frac{0.45\ in}{2} = 8.15x10^{-6}\ in$$

$$\delta_y = \epsilon_y\ \frac{L_y}{2} = 1.20x10^{-5}\ \frac{0.45\ in}{2} = 2.70x10^{-6}\ in$$

(5-28)

$$\delta_z = \max\left( \frac{L_x^2}{16} \mathcal{C}_x , \frac{L_y^2}{16} \mathcal{C}_y \right)$$

(5-20)

$$\delta_z = \max\left( \frac{(0.45in)^2}{16} 1.01x10^{-3}in^{-1} , \frac{(0.45in)^2}{16} 3.58x10^{-4}in^{-1} \right) = 1.28x10^{-5} \ in$$

$$F_{ax} = K_a\delta_x + K_{a}\delta_y + K_{a}\delta_z$$

(5-24)

$$F_{ax} = 20,868\frac{lbf}{in}8.15x10^{-6} \ in + 0 + 24,821\frac{lbf}{in}1.28x10^{-5} \ in = 0.488 \ lbf$$

$$F_{ay} = K_a\delta_x + K_a\delta_y + K_{a}\delta_z$$

(5-25)

$$F_{ay} = 0 + 20,868\frac{lb}{in}2.70x10^{-6} \ in + 24,821\frac{lb}{in}1.28x10^{-5} \ in = 0.374 \ lbf$$

$$F_{ty} = K_t\delta_x + K_{a}\delta_y + K_{t}\delta_z$$

(5-27)

$$F_{ty} = 13,065\frac{lbf}{in}8.15x10^{-6}in + 0 + 0 = 0.106 \ lbf$$

$$F_{tx} = K_a\delta_x + K_t\delta_y + K_{t}\delta_z$$

(5-26)

$$F_{tx} = 0 + 13,065\frac{lbf}{in}2.70x10^{-6} \ in + 0 = 0.0353 \ lbf$$

$$F_{ip} = \max\left( \sqrt{F_{ax}^2 + F_{tx}^2} , \sqrt{F_{ty}^2 + F_{ay}^2} \right)$$

(5-23)

$$F_{ip} = \max\left( \sqrt{(0.488)^2 + (0.0353)^2} , \sqrt{(0.106)^2 + (0.374)^2} \right)lbf = 0.489 \ lbf$$

$$\tau_{ip} = \frac{F_{ip}}{A} = \frac{0.489 \ lbf}{8.51x10^{-4} \ in^2} = 575 \ psi$$

(5-22)

$$F_{vx} = K_v \ \delta_z + K_{va}\delta_x + K_{t}\delta_y$$

(5-30)

$$F_{vx} = 57,517\frac{lbf}{in} \ 1.28x10^{-5} \ in + 24,821\frac{lbf}{in}8.15x10^{-6} \ in + 0 = 0.939 \ lbf$$

$$F_{vy} = K_v \ \delta_z + K_t\delta_x + K_{va}\delta_y$$

(5-31)

$$F_{vy} = 57,517\frac{lbf}{in} \ 1.28x10^{-5} \ in + 0 + 24,821\frac{lbf}{in}2.70x10^{-6} \ in = 0.803 \ lbf$$

$$\sigma_z = \max\left( \frac{F_{vx}}{A}, \frac{F_{vy}}{A} \right) = \max\left( \frac{0.939 lbf}{8.51x10^{-4}in^2}, \frac{0.803 lbf}{8.51x10^{-4}in^2} \right) = 1,103 \ psi$$

(5-29)

$$\tau = \sqrt{\tau_{ip}^2 + \sigma_z^2} = \sqrt{(575psi)^2 + (1103psi)^2} = \mathbf{1,244 \ psi}$$

(5-21)

## Uncorrected Vibration Life

$\tau$ = *Solder Shear Stress* = 1,244 *psi*

From Table 5-3:

$K'$ = *Solder Vibration Life Multiplier* = 233 *MPa* = 33,793 *psi*

$\lambda$ = *Solder Vibration Life Exponent* = 4.00

$$n_f = \frac{(K')^\lambda}{(0.683 + 0.271(2)^\lambda + 0.0456(3)^\lambda)\ \tau^\lambda}$$

$$n_f = \frac{(33,793\ psi)^4}{(0.683 + 0.271(2)^4 + 0.0456(3)^4)\ (1,244\ psi)^4} = \mathbf{6.25x10^4\ cycles}$$

(5-32)

## Reliability Correction Factor

$F$ = *Cumulative Failure Probaility* = 1% = 0.01

From Table 5-4 for vibration of a leadless chip-carrier:

$a'$ = *Weibull Scale Parameter* = 1.130

$\beta$ = *Weibull Shape Parameter* = 3.00

$$K_R = a'\ (\text{-}\ln(1-F))^{\frac{1}{\beta}} = 1.130\ (\text{-}\ln(1-0.01))^{\frac{1}{3.00}} = \mathbf{0.244}$$

(5-36)

## Corrected Vibration Life

The corrected vibration life for the leaded chip-carrier on the module described above is determined to be 2.4 minutes by applying the reliability correction factor to the uncorrected life as follows:

$n_f$ = *Uncorrected Cycles to Fail* = 6.25x10⁴ *cycles*

$K_R$ = *Reliability Correction Factor* = 0.244

$f_n$ = *Dominant Resonant Frequency* = 107 *Hz*

$$N_F = K_R\ n_f = 0.244\ x\ 6.25x10^4\ cycles = 1.53x10^4\ cycles$$

(5-33)

$$T_F = \frac{N_F}{f_n} = \frac{1.53x10^4\ cycles}{107\ Hz} = \mathbf{143\ sec} = \mathbf{2.4\ min}$$

(5-34)

Once again the very stiff lead actually reduces the thermal cycling life because there is a much smaller fillet than is the case with a leadless chip-carrier.

# 5.7  FACTORS IN RECENT DEVELOPMENTS

Although the principles described in this chapter are applicable to a wide range of electronic packaging configurations, special consideration may need to be given for life analysis of recent electronic packaging configurations (see 1.1.3). Since relative life analysis is independent of the specific packaging configuration, no special consideration is required if the proper fatigue slope is used for the relevant failure mode. As before, the guidelines and assumptions described for the appropriate conventional electronic packaging analyses apply.

## Ball-Grid Arrays (see Figure 1-5)

Solder life analysis of BGAs without underfill (see Figure 5-5) is determined by using the appropriate solder strain factor (see Table 5-2) for the BGA configuration. If the solder is other than tin-lead eutectic, different solder fatigue exponents and/or coefficients may be required.

Ball grid arrays (and other components) with underfill would typically require a finite-element model to determine the stress and/or strain in the solder joint. For thermal cycling situations, this finite-element model would probably need to be non-linear to account for the yielding that characterizes low-cycle fatigue. For configurations with a large number of solder balls, substructuring (see 4.3.1) may be required [49]. Once the stresses/strains in the solder are determined, the equations described in section 5.2 may be used.

## Multichip Modules (see Figure 1-6)

Life analysis of multichip modules would typically be based upon applying the stresses and strains determined from a finite-element model (see 4.3.2 and 4.6.3) to the appropriate low or high cycle fatigue relationships (see 5.3.1).

## Flip-Chips (see Figure 1-7)

Flip-chip configurations with beam leads or solder bumps with underfill would typically use a finite-element model (see 4.3.2 and 4.6.3) to determine stresses and strains, and the resulting life of individual parts (see 5.2 and 5.3.1).

Configurations with solder bumps that do not include underfill may be analyzed in a manner similar to BGAs (see Figure 5-5) using the BGA strain factor (see Table 5-2). The appropriate solder fatigue exponents and/or coefficients for the solder used must be considered.

## Chip-on-Board (see Figure 1-8)

Because of the effect of the encapsulant, chip-on-board configurations would typically require a finite-element model to determine the stress and/or strain in the packaging configuration. If yielding is present, a non-linear finite-element model is required. Because the geometry of the encapsulant can significantly influence stresses/strains, it is important to properly determine and/or model the encapsulant configuration (especially that between the die and PWB). Once the stresses/strains are determined, the equations described in section 5.2 and 5.3.1 may be used to determine life.

## 5.8    VERIFICATION

Any results from life analysis should be verified to help avoid any inaccuracies that might arise. This verification may be by testing, use of a simplified model, or other methods as appropriate. For information on verification, see Chapter 8.

## 5.9    LIFE ANALYSIS CHECKLIST

### 5.9.1    Applicable to All Life Analysis

☐   Has the input environment been properly defined?

☐   Has all source material (material properties, environmental data, etc.) been verified against the references?

### 5.9.2    Applicable to Vibration Life Analysis

☐   Has proper distinction been made for the use of the scaling factors K (1,000) and M (1,000,000) in the material properties?

### 5.9.3    Applicable to Random Vibration Life

☐   Has the correct frequency been used to relate vibration duration to cycles applied?

### 5.9.4    Applicable to Solder Vibration

☐   Has the PWB curvature been calculated properly?

### 5.9.5    Applicable to Solder Thermal Cycling

☐   Has the CTE variation over the surface of the PWB been properly considered?

### 5.9.6    Applicable to Leadless Component Analysis

☐   Do the solder joint geometry parameters correspond to actual solder joint geometry?

### 5.9.7    Applicable to Leaded Component Analysis

☐  Have the lead stiffnesses been properly calculated?

☐  Has the solder joint area and height been properly calculated?

☐  Have the effects of combined thermal cycling and vibration been considered?

### 5.9.8    Applicable to Recent Developments

☐  Has the correct fatigue exponent and/or coefficient been used for solder other than tin-lead eutectic?

☐  Has a non-linear finite-element model been used for low-cycle fatigue situations?

### 5.9.9    Environmental Data Required

The following environmental data is typically required for various analyses. Typical units are provided for convenience, and care must be taken to ensure that the system of units is consistent with the material properties and model dimensions. In many cases, these items are based upon the results of mechanical analysis, so additional environmental data is required to perform the analysis (see 4.11.6).

**Generic Vibration**

☐  Stress in component (MPa, psi, etc.)

**Solder Vibration**

☐  PWB surface strain (dimensionless)

☐  PWB curvature (1/m, 1/cm, 1/in, etc.)

**Generic Thermal Cycling**

☐  Thermal cycling range (°C, °F, etc.)

☐  Component temperature rise (°C, °F, etc.)

**Solder Thermal Cycling**

☐  Dwell time (min)

☐  Average solder temperature (°C)

**Vibration Relative Life**

☐   Power spectral density (for random vibration, $g^2/Hz$)

☐   Acceleration level (for sine vibration and shock, g)

## 5.9.10   Material Properties Required

The following material properties are typically required for various analyses. Typical units are provided for convenience, and care must be taken to ensure that the system of units is consistent for any analysis. Although some material properties have been provided in this chapter, the reader is strongly encouraged to independently verify these values. In addition to the properties listed below, additional information is required if lead stiffnesses or similar parameters need to be calculated (see 4.11.7).

**Solder Vibration**

☐   Solder shear modulus (MPa, psi, etc.)

☐   Solder vibration life multiplier (MPa, psi, etc.)

☐   Solder vibration life exponent (dimensionless)

☐   Weibull parameters (dimensionless)

**Solder Thermal Cycling**

☐   Coefficient of thermal expansion (ppm/°C, in/in/°C, in/in/°F, etc.)

☐   Solder fatigue ductility coefficient (dimensionless)

☐   Solder fatigue ductility exponent (dimensionless)

☐   Life correction cycles (cycles)

☐   Life correction exponent (dimensionless)

☐   Weibull parameters (dimensionless)

**Mechanical Hardware Life Analysis**

☐   Fatigue multiplier (MPa, psi, etc.)

☐   Fatigue exponent (dimensionless)

**Plated-Through Hole life**

☐   Coefficient of thermal expansion (ppm/°C, in/in/°C, in/in/°F, etc.)

☐   Modulus of elasticity of plating (MPa, psi, etc.)

☐   Modulus of plasticity of plating (MPa, psi, etc.)

☐   Yield stress of plating (MPa, psi, etc.)

☐   Modulus of elasticity of PWB (MPa, psi, etc.)

☐   Plating ductility (dimensionless)

☐   Plating ultimate strength (MPa, psi, etc.)

**Combined Thermal/Vibration**

☐   Yield stress of lead (MPa, psi, etc.)

# 5.10   REFERENCES

[33]   Engelmaier, W.; "Surface Mount Solder Joint Long-Term Reliability: Design, Testing, Prediction", IPC-TP-797; Institute for Interconnecting and Packaging Electronic Circuits, Lincolnwood, IL

[34]   Vaynman, S.; "Energy-Based Methodology for Solder Fatigue-Life Prediction", Final Report; BIRL Industrial Research Laboratory, Northwestern University, Evanston, IL, 1991

[35]   Vaynman, S.; "Energy Based Methodology for Solder Fatigue-Life Prediction", Project A279; BIRL, Industrial Research Laboratory, Northwestern University, Evanston, IL, 1992

[36]   IPC; "Guidelines for Accelerated Reliability Testing of Surface Mount Solder Attachments", IPC-SM-785; Institute for Interconnecting and Packaging Electronic Circuits, Lincolnwood, IL, 1992

[37]   Solomon, Brzozowski, and Thompson; "Prediction of Solder Joint Fatigue Life", 88CRD101; GE Corporate Research and Development, Schenectady, NY, 1988

[38]   McKeown, S.; "Solder Life Prediction for Leadless Chip Carriers", Proceedings of the 11th Digital Avionics Systems Conference, October 5-8, 1992, Seattle, WA; IEEE, New York, NY

[39]   Wild, R.; "Solder Properties and Characteristics", presented at the Second Annual Soldering Technology for Electronics Conference, June 7-8, 1988, Binghamton University, Binghamton, NY

[40]   Nelson, W.; "Hazard Plotting for Incomplete Failure Data", Journal of Quality Technology, Volume 1, Number 1, January 1969, pp 27-51

[41] Finke and Heberling; "Determination of Thermal-Expansion Characteristics of Metals Using Strain Gages", Experimental Mechanics, April 1978, pp 155-158

[42] Measurements Group; "How to Measure Expansion Coefficients with Strain Gages", R & E Technical Education Newsletter, No. 32, February 1981; Measurements Group, Raleigh, NC

[43] Steinberg, D.; "Preventing Thermal Cycling and Vibration Failures in Electronic Equipment", presented August 23, 1989 at McDonnell Douglas; St. Louis, MO, 1989

[44] McKeown, S.; "Solder Life Prediction of Leadless and Leaded Surface Mount Components under Thermal Cycling and Vibration"; (presented at the 1993 ASME International Electronics Packaging Conference, Binghamton, NY, September 29-October 2, 1993); Advances in Electronic Packaging - 1993, American Society of Mechanical Engineers, New York, NY, 1993; EEP-Vol. 4-2; Volume 2, pp 987-994

[45] Dowling, N.; "Fatigue Failure Predictions for Complicated Stress-Strain Histories"; Journal of Materials; Volume 7, No. 1, 1972; pp V-1 to V-17

[46] Engelmaier, W; "Enviromental Stress Screening and use Environments - Their Impact on Solder Joint and Plated-Through Hole Reliability"; Engelmaier Associates, Mendham, NJ; IEPS, Proceedings of the Technical Conference, 1990 International Electronics Packaging Conference, Marlborough, MA. September 10-12, 1990; International Electronics Packaging Society, 114 North Hale Street, Wheaton IL; pp 391-392

[47] Shigley and Mischke; Mechanical Engineering Design, Fifth Edition; McGraw-Hill, 1989

[48] Lapedes, D. (ed.); McGraw-Hill Dictionary of Scientific and Technical Terms, Second Edition; McGraw-Hill, 1978

[49] Cheng, Chiang, and Lee; "An Effective Approach for Three-Dimensional Finite Element Analysis of Ball Grid Array Typed Packages"; Journal of Electronic Packaging, Volume 120, Number 2; American Society of Mechanical Engineers, June 1998; pages 129-134

# Chapter 6
# Other Analysis

## 6.1   BACKGROUND

Many of the needs that arise with electronic packaging can be categorized as thermal, mechanical, or life analysis and may be analyzed as described in the previous chapters. In some cases, situations arise that do not neatly fit into traditional thermal or mechanical analysis, and a special approach is required. These situations requiring a special approach include:

- Fire-resistance analysis
- Pressure transducer rupture
- Humidity analysis
- Pressure-drop analysis

In other cases, alternative analytical approaches can be used to simplify the required calculations or provide a method of verification. These approaches include:

- Similarity Considerations
- Energy-Based Methods

## 6.2   FIRE-RESISTANCE ANALYSIS

Some electronic hardware, particularly that which is designed for mounting on aircraft engines, is required to pass a fire-resistance test. Typically a flame is specified that provides a certain amount of heat into a calibration tube, and the electronic unit is required to withstand exposure to the flame without becoming detached from its mounting, or leaking any fluid (see Figure 6-1).

### 6.2.1   Approach

Fire-resistance analysis is conducted as follows:

- Determine the convection coefficient between the chassis and the hot gas based upon the heat input to the calibration tube and changes in geometry between the calibration tube and the electronic chassis
- Determine the performance of the electronic chassis in the flame based upon the convection input from the flame and other appropriate boundary conditions (such as internal cooling and internal power dissipation) and transient thermal effects.

219

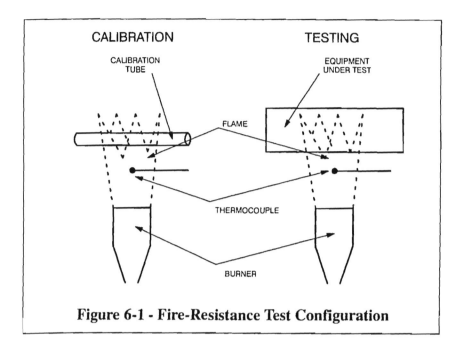

**Figure 6-1 - Fire-Resistance Test Configuration**

## 6.2.2    Determination of Input Convection

The convection input surrounding the chassis is determined by using the heat input into the calibration tube to determine the effective fluid velocity. This velocity is then used to calculate the convection coefficient between the hot gas and the chassis.

**Combustion Processes [51]**

For combustion of a typical hydrocarbon with excess air, the chemical equation is expressed as follows:

$$C_yH_z + (1 + x)\left(\frac{z}{4} + y\right)(O_2 + 3.76N_2) \rightarrow$$

$$yCO_2 + \frac{z}{2}H_2O + x\left(\frac{z}{4} + y\right)O_2 + 3.76(1 + x)\left(\frac{z}{4} + y\right)N_2$$

$$(6\text{-}1)$$

$y$ = *Number of Carbon Atoms in Molecule*

$z$ = *Number of Hydrogen Atoms in Molecule*

$x$ = *Fraction of Excess Air*

In the above combustion relationship, the excess air serves to limit the temperature of the flame. As excess air is increased, the flame temperature is lowered. Figure 6-2 illustrates the relationship between flame temperature and excess air for some typical hydrocarbon fuels. Another effect of excess air is that the composition of the products becomes closer to pure air since the products are diluted by the air (see Figure 6-3).

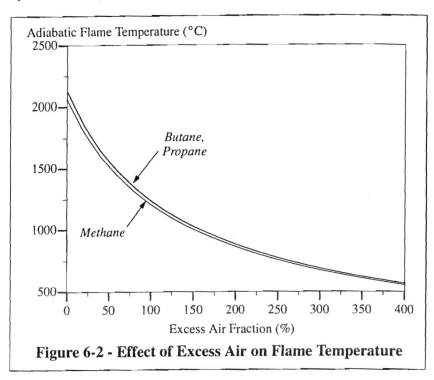

**Figure 6-2 - Effect of Excess Air on Flame Temperature**

## Properties of Mixtures of Gases

Thermal evaluation of the hot gas requires obtaining the properties of the products of combustion. The specific heat and density of the gas may be obtained by using a "mixture rule" (where the properties of each component are weighted based upon the fraction of that component). Viscosity and thermal conductivity relationships are more complex and may not be obtained by the mixture rule. To obtain the temperatures required for typical fire-resistance tests, excess air is required which will tend to dilute the combustion products and make the gas properties close to that of air. For simplicity it is assumed that the properties of the flame are identical to that of air at the same temperature. Since the thermal performance of the gas is based

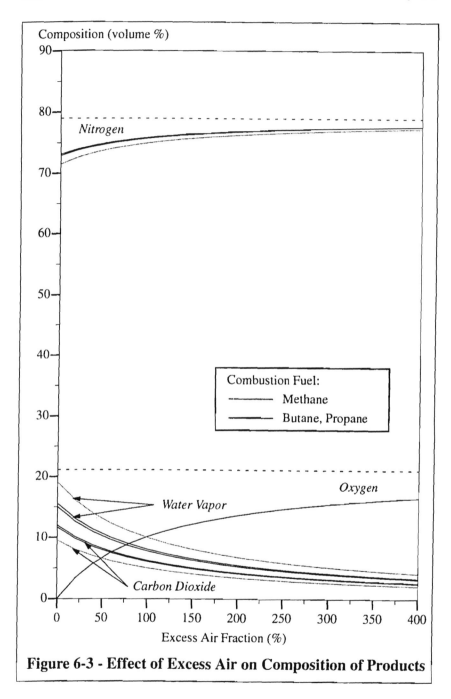

**Figure 6-3 - Effect of Excess Air on Composition of Products**

upon heat input to a calibration tube, variations in gas properties due to composition tend to cancel each other (depending upon the exponents of the Reynolds and Prandtl numbers in various flow regimes). Specific properties for products of combustion, if required, may be obtained from testing or a literature search.

**Determination of Fluid Velocity**

The fluid velocity of the flame is determined as follows:

1. Use the temperature difference, heat flow, and surface area of calibration tube to determine the convection coefficient

2. Combine the convection coefficient with the thermal conductivity of the hot gas and the diameter of the calibration tube to determine the Nusselt number.

3. Use the Nusselt number in conjunction with the relationship for heat transfer over a tube and the Prandtl number of the hot gas to determine the Reynolds number.

4. Use the Reynolds number in conjunction with the tube diameter, and viscosity of the hot gas to determine the effective gas velocity.

Convection Coefficient over Calibration Tube

The convection coefficient over the calibration tube is determined as follows:

$$h_t = \frac{Q}{\pi \, d \, l \, \Delta T} \tag{6-2}$$

$$\Delta T = T_f - T_t \tag{6-3}$$

$h_t = $ *Convection Coefficient over Calibration Tube*

$Q = $ *Heat Transfer Rate to Calibration Tube*

$d = $ *Diameter of Calibration Tube*

$l = $ *Length of Calibration Tube Exposed to Flame*

$\Delta T = $ *Temperature Difference between Tube and Flame*

$T_f = $ *Flame Temperature*

$T_t = $ *Tube Temperature*

Nusselt Number for Calibration Tube

The Nusselt number corresponding to the above convection coefficient for the calibration tube is determined as follows:

$$Nu_t = \frac{h_t \, d}{k_f} \qquad (6-4)$$

$Nu_t$ = Nusselt Number for Calibration Tube
$h_t$ = Convection Coefficient over Calibration Tube
$d$ = Diameter of Calibration Tube
$k_f$ = Thermal Conductivity of Flame

Reynolds Number for Calibration Tube

The Reynolds number for flow over a calibration tube is determined by using the relationship for the Nusselt number of a cylinder in crossflow [53]:

$$Nu_t = C \, Re_t^{\,m} \, Pr^{\frac{1}{3}} \qquad (6-5)$$

Solving for the Reynolds number:

$$Re_t = \left(\frac{Nu_t}{C \, Pr^{\frac{1}{3}}}\right)^{\frac{1}{m}} \qquad (6-6)$$

$Re_t$ = Reynolds Number over Calibration Tube
$Nu_t$ = Nusselt Number for Calibration Tube
$Pr$ = Prandtl Number
$C, \; m$ = Constants (see text and Table 6-1)

Typical values for the constants used in equation 6-6 for forced convection over a cylinder in crossflow are listed in Table 6-1. The characteristic dimension for these relationships is the tube diameter.

| Table 6-1 - Equation 6-6 Constants for Cylinder in Crossflow [53] | | |
|---|---|---|
| $\dfrac{Nu_t}{Pr^{\frac{1}{3}}}$ | C | m |
| 3.8 – 32.5 | 0.683 | 0.466 |
| 32.5 – 135 | 0.193 | 0.618 |

Fluid Velocity

The effective velocity of the hot gas is determined as follows:

$$\text{Re}_t = \frac{d\,V_e}{v} = \frac{\rho\,d\,V_e}{\mu} \tag{6-7}$$

Solving for the effective velocity:

$$V_e = \frac{v\,\text{Re}_t}{d} = \frac{\mu\,\text{Re}_t}{\rho\,d} \tag{6-8}$$

$$v = \frac{\mu}{\rho} \tag{6-9}$$

$\text{Re}_t$ = *Reynolds Number over Calibration Tube*
$d$ = *Tube Diameter*
$V_e$ = *Effective Velocity of Hot Gas*
$v$ = *Kinematic Viscosity*
$\mu$ = *Absolute Viscosity*
$\rho$ = *Density of Hot Gas*

## Determination of Convection to Chassis

Reynolds Number

The Reynolds number is determined using the same fluid properties used for the above velocity calculation, but using the distance parallel to the flow as the characteristic dimension.

$$\text{Re}_c = \frac{l\,V_e}{v} = \frac{\rho\,l\,V_e}{\mu} \tag{6-10}$$

$\text{Re}_c$ = *Reynolds Number over Chassis*
$l$ = *Length of Chassis Parallel to Flow*
$V_e$ = *Effective Velocity of Hot Gas*
$v$ = *Kinematic Viscosity*
$\mu$ = *Absolute Viscocity*
$\rho$ = *Density of Hot Gas*

Nusselt Number

The Nusselt number for flow over the chassis is determined by using equation 3-22 with the characteristic dimension of length parallel to flow.

$$Nu_c = C \, Re_c^{\,m} Pr^n \qquad (3\text{-}22)$$

$Nu_c$ = Nusselt Number over Chassis
$Re_c$ = Reynolds Number over Chassis
$Pr$ = Prandtl Number
$C, \, m, \, n$ = Constants (see text and Table 6-2)

Typical values for the constants used in equation 3-22 for forced convection over the surface of an electronic chassis are listed in Table 6-2. Increased viscosity of air at high temperatures makes laminar flow possible at higher fluid velocities. Therefore, Table 6-2 also includes the appropriate constants for laminar flow.

| Table 6-2 - Equation 3-22 Constants for Convection of a Gas Over a Chassis | | | |
|---|---|---|---|
| Re | C | m | n |
| <500,000 [53] | 0.664 | 0.5 | 0.333 |
| >500,000 | 0.0369 | 0.8 | 0.6 |

Convection to Chassis

The convection to the chassis is determined by solving the definition of the Nusselt number for the convection coefficient as follows:

$$Nu_c = \frac{h_c \, l}{k_f} \qquad (6\text{-}11)$$

$$h_c = \frac{Nu_c \, k_f}{l} \qquad (6\text{-}12)$$

$Nu_c$ = Nusselt Number over Chassis
$h_c$ = Convection Coefficient over Chassis
$l$ = Length Parallel to Flow
$k_f$ = Thermal Conductivity of Flame

**Underlying Assumptions**

- Properties of the hot gas are the same as for air at the same temperature

- Radiation to the hot gas is ignored

- The Nusselt number for flow over a flat plate is the same whether the plate is heated or cooled by the gas

## 6.2.3    Fire-Resistance Performance

**Analysis**

The fire-resistance performance is determined by developing a thermal model of the chassis (see 3.6). Since the flame imparts a high heat flux to the chassis, considerable temperature variations are expected over the surface. These temperature variations, generally require the use of finite-element analysis (see Figure 6-4) to properly determine the surface temperatures. Once the temperatures are determined, subsequent determinations may include the following:

- Melting of chassis walls in area of flame

- Ability of cooling paths to withstand the appropriate pressures at temperature

- Melting of mounting feet

- Temperatures of electronic components

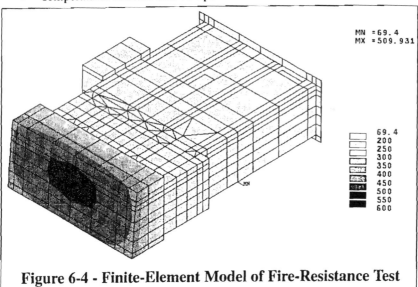

MN =69.4
MX =509.931

69.4
200
250
300
350
400
450
500
550
600

## Figure 6-4 - Finite-Element Model of Fire-Resistance Test

**Underlying Assumptions**

- Melting effects (heat of fusion) on thermal performance is neglected

## 6.2.4    Typical Fire-Resistance Analysis

For the electronic chassis described in 3.7.4 (see Figures 3-17 and 6-5) exposed to a 2,000°F flame impinging on the front and top, determine the following:

- Convection coefficient between chassis and flame

- Heat input to chassis operating at an average temperature of 200°F

- Approximate temperature rise of the chassis after 5 minutes of exposure to the flame based upon a chassis weight of 20 pounds

The flame has sufficient intensity to produce an input of 4,500 Btu/hour into a 15 in calibration tube at 200°F with a diameter of 0.5 in [50].

**Convection Coefficient**

Mixture Composition

For a temperature of 2,000°F (1,093°C) using methane as a fuel, 126% excess air is required (see Figure 6-2). Composition of the products (see Figure 6-3) is 4% carbon dioxide, 9% water, 11% oxygen, and 75% nitrogen which is somewhat close to the composition of dry air (21% oxygen and 79% nitrogen). The sensitivity to variation in properties from the assumed air will be determined later.

Convection to Calibration Tube

The convection coefficient over the calibration tube is determined:

$Q$ = *Heat Transfer Rate to Calibration Tube* = $4,500 \dfrac{Btu}{hour}$

$d$ = *Diameter of Calibration Tube* = $0.5\ in$

$l$ = *Length of Calibration Tube Exposed to Flame* = $15\ in$

$T_f$ = *Flame Temperature* = $2,000°F$

$T_t$ = *Tube Temperature* = $200°F$

$$\Delta T = T_f - T_t = 2,000°F - 200°F = 1,800°F \qquad (6\text{-}3)$$

$$h_t = \frac{Q}{\pi\ d\ l\ \Delta T} = \frac{4,500\frac{Btu}{hour}}{\pi x 0.5 in x 15 in x 1,800°F}\left(\frac{12 in}{ft}\right)^2 = 15.3\frac{Btu}{hr\text{-}ft^2\text{-}°F} \qquad (6\text{-}2)$$

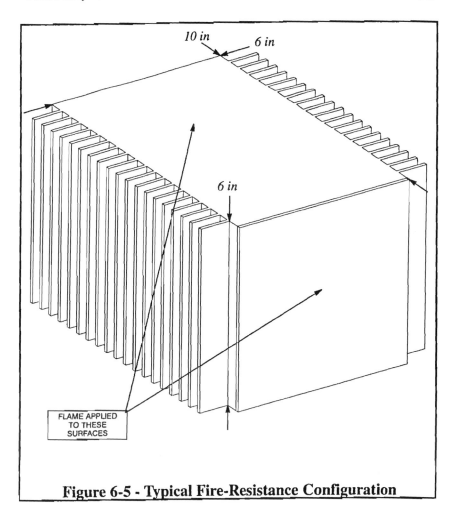

**Figure 6-5 - Typical Fire-Resistance Configuration**

<u>Nusselt and Reynolds Numbers for Calibration Tube</u>

The Nusselt number corresponding to the above convection coefficient is determined as follows:

$$h_t = Convection\ Coefficient\ over\ Calibration\ Tube\ =\ 15.3\frac{Btu}{hr\text{-}ft^2\text{-}°F}$$
$$d = Diameter\ of\ Calibration\ Tube\ =\ 0.5\ in$$

For air at film temperature of $1,100°F$ $(2,000°F + 200°F)/2$ [63]:

$k_f = Thermal\ Conductivity\ of\ Flame = 0.035 \dfrac{Btu}{hr\text{-}ft\text{-}°F}$

$Pr = Prandtl\ Number = 0.70$

$$Nu_t = \frac{h, d}{k_f} = \frac{15.3 \frac{Btu}{hr\text{-}ft^2\text{-}°F} x 0.5 inx \frac{ft}{12in}}{0.035 \frac{Btu}{hr\text{-}ft\text{-}°F}} = 18.2$$

$$\frac{Nu_t}{Pr^{\frac{1}{3}}} = \frac{18.2}{0.7^{\frac{1}{3}}} = 20.5$$

(6-4)

From Table 6-1 for $Nu/Pr^{0.333} = 20.5$:

$C = 0.683$

$m = 0.466$

$$Re_t = \left(\frac{Nu_t}{C\,Pr^{\frac{1}{3}}}\right)^{\frac{1}{m}} = \left(\frac{18.2}{0.683 x 0.7^{\frac{1}{3}}}\right)^{\frac{1}{0.466}} = 1,480$$

(6-6)

## Fluid Velocity

The effective hot gas velocity is determined from the Reynolds number:

$Re_t = Reynolds\ Number\ over\ Calibration\ Tube = 1,480$

$d = Tube\ Diameter = 0.5\ in$

For air ([52],[63]) at film temperature of $1,100°F$ $(1,559°R)$:

$\mu = Absolute\ Viscocity = 0.093 \dfrac{lbm}{hr\text{-}ft}$

$R = Universal\ Gas\ Constant = 53.35 \dfrac{ft\text{-}lbf}{lbm\text{-}°R}$

$$\rho_0 = Hot\ Gas\ Density = \frac{p}{RT} = \frac{14.7psi x 144 \frac{in^2}{ft^2}}{53.35 \frac{ft\text{-}lbf}{lbm\text{-}°R} x 1,559°R} = 0.0255 \frac{lbm}{ft^3}$$

$$V_e = \frac{\mu\ Re_t}{\rho\ d} = \frac{0.093 \frac{lbm}{hr\text{-}ft} x \frac{hr}{3,600sec} x 1,480}{0.0255 \frac{lbm}{ft^3} x 0.5 inx \frac{ft}{12in}} = 35.9 \frac{ft}{sec}$$

(6-8)

## Reynolds Number for Chassis

The Reynolds number for flow over the chassis is determined using the length parallel to flow as the characteristic dimension:

$l$ = *Length of Chassis Parallel to Flow* = 6 *in (front)*, 10 *in (top)*

$V_e$ = *Effective Velocity of Hot Gas* = $35.9 \frac{ft}{sec}$

Using properties determined before for air at film temperature of 1,100°F (1,559°R):

$\mu$ = *Absolute Viscocity* = $0.093 \frac{lbm}{hr\text{-}ft}$

$\rho_0$ = *Hot Gas Density* = $0.0255 \frac{lbm}{ft^3}$

Reynolds number for front:

$$\text{Re}_c = \frac{\rho \, l \, V_e}{\mu} = \frac{0.0255 \frac{lbm}{ft^3} x6 inx \frac{ft}{12in} x35.9 \frac{ft}{sec}}{.093 \frac{lbm}{hr\text{-}ft} x \frac{hr}{3,600sec}} = 17,718 \qquad \text{(6-10)}$$

Reynolds number for top:

$$\text{Re}_c = \frac{\rho \, l \, V_e}{\mu} = \frac{0.0255 \frac{lbm}{ft^3} x10 inx \frac{ft}{12in} x35.9 \frac{ft}{sec}}{.093 \frac{lbm}{hr\text{-}ft} x \frac{hr}{3,600sec}} = 29,531 \qquad \text{(6-10)}$$

Nusselt Number for Chassis

The Nusselt number for flow over the chassis is determined from the Reynolds number:

$Re_c$ = *Reynolds Number over Chassis* = 17,718 *(front)*, 29,531 *(top)*

For air at film temperature of 1,100°F (1,559°R) [63]:

Pr = *Prandtl Number* = 0.70

From Table 6-2 for Reynolds numbers of 17,718 and 29,531:

$C$ = 0.664

$m$ = 0.5

$n$ = 0.333

Nusselt number for front:

$$Nu_c = C \, Re_c^m Pr^n = 0.664 x 17,718^{0.5} x 0.70^{0.333} = 78.4 \qquad \text{(3-22)}$$

Nusselt number for top:

$$Nu_c = C \, Re_c^m Pr^n = 0.664 x 29,531^{0.5} x 0.70^{0.333} = 101.3 \qquad \text{(3-22)}$$

| Table 6-3 - Effect of Property Variations on Convection Coefficient | | |
|---|---|---|
| A +10% change in this Property | Produces this Change in Convection | |
| | Front | Top |
| Thermal conductivity | -0.7% | -0.7% |
| Prandtl number | -0.2% | -0.2% |
| Absolute viscosity | 0% | 0% |
| Density | 0% | 0% |

<u>Convection to Chassis</u>

The Nusselt number is used to determine the convection to the chassis:

$Nu_c$ = Nusselt Number over Chassis = 78.4 *(front)*, 101.3 *(top)*
$l$ = Length of Chassis Parallel to Flow = 6 in *(front)*, 10 in *(top)*

For air at film temperature of 1,100°F (1,559°R) [63]:

$$k_f = Thermal\ Conductivity\ of\ Flame = 0.035 \frac{Btu}{hr\text{-}ft\text{-}°F}$$

Convection coefficient for front:

$$h_c = \frac{Nu_c\ k_f}{l} = \frac{78.4 x 0.035 \frac{Btu}{hr\text{-}ft\text{-}°F}}{6inx \frac{ft}{12in}} = 5.48 \frac{Btu}{hr\text{-}ft^2\text{-}°F} \tag{6-12}$$

Convection coefficient for front:

$$h_c = \frac{Nu_c\ k_f}{l} = \frac{101.3 x 0.035 \frac{Btu}{hr\text{-}ft\text{-}°F}}{10inx \frac{ft}{12in}} = 4.25 \frac{Btu}{hr\text{-}ft^2\text{-}°F} \tag{6-12}$$

The convection between the hot gas and the chassis is *5.48 Btu/(hr-ft²-°F)* on the front of the chassis and *4.25 Btu/(hr-ft²-°F)* along the top.

<u>Sensitivity to Property Variations</u>

The effect of property variations on the convection coefficients determined above can be determined by applying a 10% change to relevant material properties and repeating the calculations to determine the change in the convection coefficient (see Table 6-3). Since the same properties are applied to both the calibration case and the analysis the effect of property variations are minor. Using air properties for the products of combustion does not significantly affect results.

**Heat Input to Chassis**

The heat input to the chassis from the flame is determined as follows:

$$T_f = Flame\ Temperature = 2,000°F$$
$$T_c = Chassis\ Temperature = 200°F$$
$$h_f = Convection\ Coefficient\ on\ Front = 5.48\frac{Btu}{hr\text{-}ft^2\text{-}°F}$$
$$h_t = Convection\ Coefficient\ on\ Top = 4.25\frac{Btu}{hr\text{-}ft^2\text{-}°F}$$
$$A_t = Area\ of\ Top = 6\ in\ x\ 10\ in = 60\ in^2$$
$$A_f = Area\ of\ Front = 6\ in\ x\ 6\ in = 36\ in^2$$

$$q_{free} = (h_s A_s + h_t A_t + h_b A_b + h_e A_e + h_{fl} A_f)\ (T_{chas} - T_\infty) \qquad (3\text{-}23)$$

Neglecting unused terms and changing subscripts:

$$q = (h_t A_t + h_f A_f)\ (T_f - T_c)$$

$$q = \left[ 4.25\frac{Btu}{hr\text{-}ft^2\text{-}°F}x60\frac{in^2}{\frac{144in^2}{ft^2}} + 5.48\frac{Btu}{hr\text{-}ft^2\text{-}°F}x\frac{36\ in^2}{\frac{144in^2}{ft^2}}\right](2,000°F - 200°F)$$

$$(6\text{-}13)$$

$$q = 5,654\frac{Btu}{hr}$$

$$q = 5,654\frac{Btu}{hr}x\frac{0.293W}{\frac{Btu}{hr}} = 1,657\ W$$

The heat input to the chassis due to the flame is therefore shown to be *1,657 Watts*. It must be noted that this power value assumes that the chassis is at a constant 200°F, which is actually not true. Any localized temperature rise of the chassis will reduce the heat transfer, so this power value represents an upper limit. If a more detailed analysis is required to characterize the local temperature rises, a finite-element model may be used with the appropriate convection coefficients previously derived.

**Transient Temperature Rise**

An approximate transient temperature rise may be determined by applying the power level determined above to the thermal mass of the chassis as follows:

$P_{tot}$ = Total Power Added = 1,657 W

$M_{chas}$ = Chassis Mass = 20 lbm

$c_{chas}$ = Chassis Specific Heat = $0.3\frac{Btu}{lbm\text{-}°F}$ (estimated)

$Mc_{chas}$ = Chassis Thermal Mass = $6\frac{Btu}{°F}$

$\Delta t$ = Time Increment = 5 min

$$Mc_{chas}\frac{dT_{chas}}{dt} = P_{tot} - \Sigma\frac{(T_i - T_{chas})}{\theta_i} - q_{rad} - q_{free} - q_{forced} \qquad (3\text{-}54)$$

Neglecting heat transfer to environment and between modules and chassis, and using discrete representation of derivative:

$$Mc_{chas}\frac{\Delta T_{chas}}{\Delta t} = P_{tot} \qquad (6\text{-}14)$$

$$\Delta T_{chas} = \frac{P_{tot}\,\Delta t}{Mc_{chas}} = \frac{1,657Wx\frac{\frac{Btu}{hr}}{0.293W}x5minx\frac{hr}{60min}}{6\frac{Btu}{°F}} = 78.6°F \qquad (6\text{-}15)$$

Therefore, a 5-minute exposure to the flame described above will increase the average chassis temperature by $79°F$. Since the additional heat lost to the ambient is neglected, and the power input may be less than calculated (as discussed previously) the actual temperature rise may be less.

## 6.3   PRESSURE TRANSDUCER RUPTURE

Some electronics are designed to be used in control applications that require sensing of pressures to perform the desired function. Some of these configurations include pressure transducers within the electronic control that sense hot, high-pressure gases. Under normal operation, the transducers are sealed so that no gas flows along the transducer tube. In this case, the only heat input to the control is by conduction through the tube wall and the stagnant gas inside the tube. However, if one of the pressure transducers were to rupture, hot gas would flow into the control resulting in increased temperature of the electronics (see Figure 6-6). Typically, an orifice is provided in the pressure path to limit flow into the control. Determining the magnitude of the increased temperature can be used in the design phase to help avoid thermal problems.

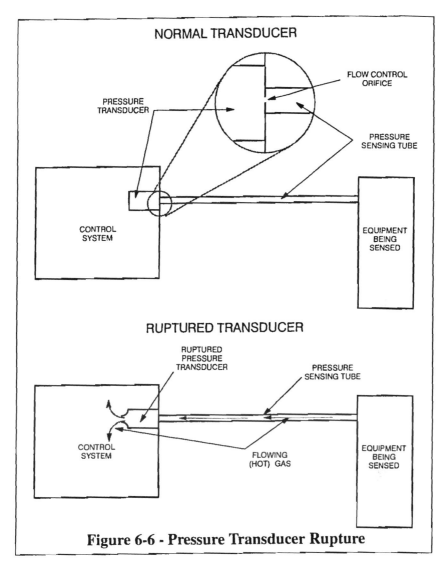

**Figure 6-6 - Pressure Transducer Rupture**

## 6.3.1    Approach

Pressure transducer rupture analysis is conducted as follows:

- Determine the flow rate of the hot gas through the limiting orifice.

- Determine the thermal performance of the electronic chassis based upon the additional heat input due to the flowing gas.

## 6.3.2    Determination of Flow Rate

**Compressible Flow of Ideal Gas in Orifice**

The maximum flow rate is determined by using gas dynamic relationships for isentropic compressible flow, an ideal gas in a converging nozzle:

$$\dot{m} = A\rho_0 \sqrt{2c_pT_0} \left(\frac{p}{p_0}\right)^{\frac{1}{\gamma}} \sqrt{1 - \left(\frac{p}{p_0}\right)^{\frac{\gamma-1}{\gamma}}} \quad (for\, p > p_{cr})$$

(6-16)

$$\dot{m} = A\rho_0 a_0 \left(\frac{p_{cr}}{p_0}\right)^{\frac{\gamma+1}{2\gamma}} \quad (for\, p \leq p_{cr})$$

$$c_p = \frac{\gamma}{\gamma - 1} R$$

(6-17)

$$p_{cr} = p_0 \left(\frac{2}{\gamma + 1}\right)^{\frac{\gamma}{\gamma-1}}$$

(6-18)

$$a_0 = \sqrt{\gamma R T_0}$$

(6-19)

$\dot{m}$ = *Mass Flow Rate*

$A$ = *Cross-Sectional Area of Orifice*

$\rho_0$ = *Density of Gas Upstream of Orifice (temperature $T_0$)*

$c_p$ = *Specific Heat of Gas at Constant Pressure*

$T_0$ = *Absolute Temperature of Gas Upstream of Orifice*

$p_0, p$ = *Presssure Upstream, Downstream of Orifice*

$\gamma$ = *Specific Heat Ratio for Gas (1.4 for air)*

$p_{cr}$ = *Critical Pressure for Gas*

$a_0$ = *Speed of Sound in Gas at Temperature $T_0$*

$R$ = *Universal Gas Constant*

If the pressure drop and heat loss in the pressure sensing tube are neglected, the upstream pressure, temperature and density ($p_0$, $T_0$, $\varrho_0$) can be considered to be equal to that of the source (worst-case assumption). It should be noted that the universal gas constant (R) is based upon mass and not moles as is sometimes used in chemistry texts.

**Underlying Assumptions**

- Pressure drop in sensing tube is neglected
- Heat loss from sensing tube and orifice is neglected

## 6.3.3    Thermal Performance

**Heat Transfer from Flowing Gas**

The heat added to the chassis due to the flowing gas is determined by using the mass flow rate of the gas determined above in conjunction with the temperature difference:

$$q = \dot{m}c_p(T_0 - T_e) \tag{6-20}$$

$q$ = *Heat Transfer Rate*
$\dot{m}$ = *Mass Flow Rate*
$c_p$ = *Specific Heat of Gas at Constant Pressure*
$T_0$ = *Temperature of Gas Upstream of Orifice*
$T_e$ = *Gas Temperature Exiting from Chassis*

The worst-case condition is determined by considering the exit gas temperature ($T_e$) to be equal to the chassis temperature.

**Effect on Chassis Temperature**

The effect of the transducer rupture on thermal performance is determined by developing a thermal model of the chassis (see 3.6). If the flowing gas transfers significant heat to the chassis, considerable temperature variations over the chassis surface are to be expected. These temperature variations, may require the use of finite-element analysis (see Figure 6-4) to properly determine the surface temperatures. If the heat transfer is moderate, a classical analysis may be appropriate. Once the temperatures are determined, other determinations may include the following:

- Temperatures of electronic components
- Orifice size required to limit flow rate and resulting temperatures

If the worst-case assumptions described herein result in unmanageable restrictions in the design process, a more detailed model may be required. Additional details may include the following:

- Heat loss and pressure drop in the sensing tube
- Actual heat transfer between hot gas and chassis to determine actual exit temperature
- Local temperature variations

**Underlying Assumptions**

- Exit gas temperature equal to chassis temperature

## 6.3.4    Effect of Orifice on Response

Addition of an orifice in the pressure sensing will restrict flow and may influence the transient response of the transducer. Dynamic response of a transducer with an orifice may be determined by treating the system as a Helmholtz resonator. The resonant frequency of a Helmholtz resonator is determined as follows [55]:

$$f_0 = \frac{a}{2\pi} \sqrt{\frac{A}{l'V}} \qquad (6\text{-}21)$$

$$a = \sqrt{\gamma RT} \qquad (6\text{-}19)$$

$$l' = l + 2Kr \qquad (6\text{-}22)$$

$f_0 = Resonant\ Frequency$
$a = Speed\ of\ Sound\ in\ Gas\ at\ Transducer\ Temperature$
$A = Cross\text{-}Sectional\ Area\ of\ Orifice$
$l' = Effective\ Length\ of\ Orifice$
$V = Volume\ of\ Pressure\ Transducer\ Cavity\ Downstream\ from\ Orifice$
$\gamma = Specific\ Heat\ Ratio\ for\ Gas\ (1.4\ for\ air)$
$R = Universal\ Gas\ Constant$
$T = Temperature\ of\ Gas\ in\ Transducer$
$l = Physical\ Length\ of\ Orifice$
$K = Orifice\ End\ Correction = 0.85\ for\ flanged\ pipe\ or\ hole\ in\ plate$
$r = Radius\ of\ Orifice$

This resonant frequency may be considered as a cut-off frequency for a low-pass filter in the pressure sensing line. Transients below this cut-off frequency will be detected but those above this frequency will be attenuated. Alteratively, the peak response time to a step change may be determined from the resonant frequency as follows (for a lightly damped system):

$$T_p \approx \frac{1}{2f_0}$$

(6-23)

$T_p$ = Time for Peak Response to Step Input
$f_0$ = Resonant Frequency

### 6.3.5    Typical Pressure Transducer Rupture Analysis

A pressure transducer in an electronic control system operating at 120°F at sea level senses air at a maximum of 500°F at 600 psi (see Figure 6-7). A 0.020-in diameter orifice is provided in a 0.030-in thick plate at the input to the pressure transducer. Determine the following:

- Maximum heat input to the chassis in the event of a pressure transducer rupture

- Effect of orifice on pressure response time based upon a transducer cavity volume of 1 in³

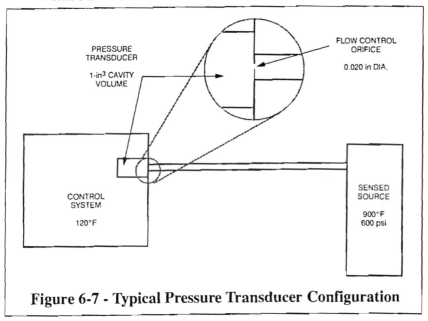

**Figure 6-7 - Typical Pressure Transducer Configuration**

**Heat Input to Chassis**

Mass Flow Rate

The maximum flow rate is determined as follows:

$A$ = Cross-Sectional Area of Orifice = $\pi \dfrac{d^2}{4} = \pi \dfrac{(0.020 in)^2}{4}$ = 0.000314 $in^2$

$T_0$ = Temperature of Gas Upstream of Orifice = 500°F = 960°R

$p_0$ = Presssure Upstream of Orifice = 600 psi

$p$ = Presssure Dowstream of Orifice = 14.7 psi

Air properties ([52],[53]) at 900°F:

$\gamma$ = Specific Heat Ratio for Gas = 1.4

$R$ = Universal Gas Constant = $53.35 \dfrac{ft\text{-}lbf}{lbm\text{-}°R} = 640.2 \dfrac{in\text{-}lbf}{lbm\text{-}°R}$

$\rho_0$ = Density of Gas Upstream of Orifice = $\dfrac{p_0}{RT_0} = 9.76 \times 10^{-4} \dfrac{lbm}{in^3}$

$$p_{cr} = p_0 \left( \frac{2}{\gamma + 1} \right)^{\frac{\gamma}{\gamma - 1}} = 600 \, psi \left( \frac{2}{1.4 + 1} \right)^{\frac{1.4}{1.4 - 1}} = 317 \, psi \qquad (6\text{-}18)$$

$$a_0 = \sqrt{\gamma R T_0} = \sqrt{1.4 \times 32.2 \frac{lbm \frac{ft}{sec^2}}{lbf} \times 53.35 \frac{ft\text{-}lbf}{lbm\text{-}°R} \times 960°R} \qquad (6\text{-}19)$$

$$a_0 = 1,519 \frac{ft}{sec}$$

$$\dot{m} = A \, \rho_0 \sqrt{2 c_p T_0} \left( \frac{p}{p_0} \right)^{\frac{1}{\gamma}} \sqrt{1 - \left( \frac{p}{p_0} \right)^{\frac{\gamma - 1}{\gamma}}} \quad (for \, p > p_{cr})$$

$$\dot{m} = A \rho_0 a_0 \left( \frac{p_{cr}}{p_0} \right)^{\frac{\gamma + 1}{2\gamma}} \quad (for \, p \leq p_{cr}) \qquad (6\text{-}16)$$

$$\dot{m} = 0.000314 \, in^2 \times 9.76 \times 10^{-4} \frac{lbm}{in^3} \times 1,519 \frac{ft}{sec} \times 12 \frac{in}{ft} \left( \frac{317 \, psi}{600 \, psi} \right)^{\frac{1.4 + 1}{2 \times 1.4}}$$

$$\dot{m} = 0.00323 \frac{lbm}{sec}$$

## Heat Transfer from Flowing Gas

The heat input to the chassis is determined as follows:

$\dot{m}$ = Mass Flow Rate = $0.00272 \dfrac{lbm}{sec}$

$T_0$ = Temperature of Gas Upstream of Orifice = 500°F

$T_e$ = Exit Gas Temperature = Chassis Temperature = 120°F

For air at 500°F [53]:

$$c_p = Specific\ Heat\ at\ Constant\ Pressure = 0.248\frac{Btu}{lbm°F}$$

For air at 120°F [53]:

$$c_p = Specific\ Heat\ at\ Constant\ Pressure = 0.241\frac{Btu}{lbm°F}$$

Average specific heat:

$$c_p = Specific\ Heat\ at\ Constant\ Pressure = 0.231\frac{Btu}{lbm°F}$$

$$q = \dot{m}c_p(T_0 - T_e) = 0.00323\frac{lbm}{sec}x0.231\frac{Btu}{lbm°F}(500°F - 120°F)$$

$$(6\text{-}20)$$

$$q = 0.284\frac{Btu}{sec}x1055\frac{J}{Btu} = 300\frac{J}{sec} = \mathbf{300\ W}$$

So the heat input to the chassis in the event of a pressure transducer rupture is *300 Watts*. The effect of this additional heat on the thermal performance of the chassis may be determined by thermal analysis as described in Chapter 3.

### Effect of Orifice on Response

The cut-off frequency and time to peak of the transducer with the orifice is determined as follows:

$l = Physical\ Length\ of\ Orifice = 0.030\ in$

$K = Orifice\ End\ Correction = 0.85\ for\ flanged\ pipe\ or\ hole\ in\ plate$

$r = Radius\ of\ Orifice = \dfrac{0.020\ in}{2} = 0.010\ in$

$A = Cross\text{-}Sectional\ Area\ of\ Orifice = \pi\dfrac{d^2}{4} = \pi\dfrac{(0.020in)^2}{4} = 0.000314\ in^2$

$V = Volume\ of\ Pressure\ Transducer\ Cavity = 1\ in^3$

$T = Temperature\ of\ Gas\ in\ Transducer = 120°F = 580°R$

Air properties [52]:

$\gamma = Specific\ Heat\ Ratio\ for\ Gas = 1.4$

$R = Universal\ Gas\ Constant = 53.35\dfrac{ft\text{-}lbf}{lbm\text{-}°R} = 640.2\dfrac{in\text{-}lbf}{lbm\text{-}°R}$

$$l' = l + 2Kr = 0.030in + 2x0.85x0.010in = 0.047\ in \qquad (6\text{-}22)$$

$$a = \sqrt{\gamma RT} = \sqrt{1.4 x 32.2 \frac{lbm \frac{ft}{sec^2}}{lbf} x 53.35 \frac{ft\text{-}lbf}{lbm\text{-}°R} x 580°R} \qquad (6\text{-}19)$$

$$a_0 = 1,181 \frac{ft}{sec}$$

$$f_0 = \frac{a}{2\pi} \sqrt{\frac{A}{l'V}} = \frac{1,181 \frac{ft}{sec} x 12 \frac{in}{ft}}{2\pi} \sqrt{\frac{0.000314 in^2}{0.047 in x 1 in^3}} = 184\,Hz \qquad (6\text{-}21)$$

$$T_p \approx \frac{1}{2f_0} = \frac{1}{2 x 184 Hz} = 0.0027\,sec = 2.7\,ms \qquad (6\text{-}23)$$

The orifice and pressure transducer described above has a cut-off frequency of *184 Hz*. Peak response to a step input will be approximately *2.7 milliseconds*.

## 6.4   HUMIDITY ANALYSIS

Humidity is an important consideration in the performance of electronic equipment. If the equipment is cooled below ambient temperature or high pressure gas is supplied to the equipment, condensed moisture may develop, possibly resulting in corrosion or electrical short circuits. Humidity is also a concern where non-hermetic components are used.

### 6.4.1   Humidity Relationships

Humidity relationships are based upon considering the water vapor and the air as a mixture of ideal gases [52]. For such a mixture, the volume fraction of each component can be represented as the ratio of partial pressures.

$$f_v = \frac{P_v}{P_{tot}} \qquad (6\text{-}24)$$

$$f_g = \frac{P_g}{P_{tot}} \qquad (6\text{-}25)$$

$$P_{tot} = P_v + P_g \qquad (6\text{-}26)$$

$f_{vg}$ = *Volume Fraction of Vapor, Gas*
$P_{vg}$ = *Partial Pressure of Vapor, Gas*
$P_{tot}$ = *Total Pressure*

## Humidity Ratio

The humidity ratio is defined as the ratio of the mass of water vapor to the mass of dry air. This ratio can be expressed in terms of partial pressures and molecular weights as follows:

$$\omega = \frac{M_v P_v}{M_g P_g} = \frac{M_v P_v}{M_g(P_{tot} - P_v)} \qquad (6\text{-}27)$$

For air and water vapor:

$$\omega = 0.622 \frac{P_v}{P_{tot} - P_v} \qquad (6\text{-}28)$$

$\omega$ = Humdity Ratio

$M_v$ = Molecular Weight of Vapor = 18.016 (for water)

$M_g$ = Molecular Weight of Gas = 28.97 (for air)

$P_{v,g}$ = Partial Pressure of Vapor, Gas

$P_{tot}$ = Total Pressure

## Relative Humidity

The relative humidity is the ratio of the volume fraction of water vapor in the mixture to the volume fraction of a saturated water vapor mixture at the same temperature. This can also be expressed in terms of partial pressures as follows:

$$\phi = \frac{P_v}{P_{sat}} \qquad (6\text{-}29)$$

Solving equation 6-28 for the vapor pressure and substituting into equation 6-29:

$$\phi = \frac{\omega \, P_{tot}}{P_{sat} \, (\omega + 0.622)} \qquad (6\text{-}30)$$

$$\log P_{sat} = A - \frac{B}{C + T} \qquad (6\text{-}31)$$

The humidity ratio (equation 6-28) can also be expressed in terms of relative humidity:

$$\omega = 0.622 \frac{\phi P_{sat}}{P_{tot} - \phi P_{sat}} \qquad (6\text{-}32)$$

$\phi$ = *Relative Humidity*

$P_v$ = *Partial Pressure of Vapor*

$P_{tot}$ = *Total Pressure*

$P_{sat}$ = *Saturation Pressure of Vapor (mmHg) [56]*

$T$ = *Temperature (°C)*

$A, B, C$ = *Constants (see Table 6-4)*

$\omega$ = *Humidity Ratio*

It must be noted that the vapor pressure equation (6-31) gives vapor pressure in millimeters of mercury from temperature in degrees Celsius. The use of other units for pressure or temperature will require conversion to maintain consistency for other humidity equations.

| Table 6-4 - Equation 6-31 and 6-35 Constants for Vapor Pressure (mmHg) vs. Temperature (°C) [56] | | | |
|---|---|---|---|
| Temperature Range (°C) | A | B | C |
| 0 – 60 | 8.10765 | 1750.286 | 235 |
| 60 – 150 | 7.96681 | 1668.21 | 228 |

**Dew Point**

The dew point of a gas vapor mixture is the temperature at which condensation occurs. For water vapor in air, the dew point is determined by solving equation 6-28 or 6-29 for vapor pressure and determining the corresponding saturation temperature:

$$P_v = \frac{\omega\, P_{tot}}{0.622 + \omega} \qquad (6\text{-}33)$$

$$P_v = \phi P_{sat} \qquad (6\text{-}34)$$

$$\log P_{sat} = A - \frac{B}{C + T} \qquad (6\text{-}31)$$

$$T_{dew} = \frac{B}{A - \log P_v} - C \qquad (6\text{-}35)$$

$P_v$ = Partial Pressure of Vapor

$P_{tot}$ = Total Pressure

$\phi$ = Relative Humidity

$\omega$ = Humidity Ratio

$P_{sat}$ = Saturation Pressure of Vapor (mmHg) [56]

$T$ = Temperature (°C)

$A, B, C$ = Constants (see Table 6-4)

$T_{dew}$ = Dew Point (°C)

It is important that the constants used in equation 6-35 be consistent with the calculated dew point.

## Underlying Assumptions

- There are no dissolved gases in either the solid or liquid phases
- The saturation pressure of the vapor phase is not influenced by the presence of the gaseous phase
- The gas and vapor are ideal gases

## 6.4.2    Typical Humidity Calculation

The air in the pressure transducer cavity described in 6.3.5 (500°F at 600 psi) was obtained by compressing sea-level air at 70°F and 80% relative humidity without adding moisture.  Determine the following:

- Relative humidity of the high-pressure air
- Dew point of the high-pressure air

### Relative Humidity of High-Pressure Air

Humidity Ratio

Since no moisture is added during the compression, the humidity ratio at sea-level and in the compressed state is the same.  Calculating the sea-level humidity ratio:

$T$ = Temperature = 70°F = 21.1°C

$\phi$ = Relative Humidity = 80% = 0.80

$P_{tot}$ = Total Pressure = 14.7 psi

From Table 6-4:

$A = 8.10765$

$B = 1750.286$

$C = 235$

$$\log P_{sat} = A - \frac{B}{C+T} = 8.10765 - \frac{1750.286}{235 + 21.1°C} = 1.273$$
(6-31)

$$P_{sat} = 10^{1.273} = 18.75 \ mmHg = 0.3626 \ psi$$

$$\omega = 0.622 \frac{\phi P_{sat}}{P_{tot} - \phi P_{sat}} = 0.622 \frac{0.80x0.3626psi}{14.7psi - 0.80x0.3626psi}$$
(6-32)

$$\omega = 0.0125 \frac{lbm \ vapor}{lbm \ dry \ air}$$

### Relative Humidity of Compressed Air

Using the same humidity ratio with the higher temperature and total pressure:

$T = Temperature = 500°F = 260°C$

$P_{tot} = Total \ Pressure = 600 \ psi$

Since the temperature of 260°C exceeds the limit of 150°C described in Table 6-4 a steam table [52] is used to determine the saturation pressure of water vapor at 500°F:

$P_{sat} = Saturation \ Pressure \ of \ Vapor = 680 \ psi$

$$\phi = \frac{\omega \ P_{tot}}{P_{sat} \ (\omega + 0.622)} = \frac{0.0125x600psi}{680psi(0.0125 + 0.622)} = 0.017 = 1.7\% \quad (6\text{-}30)$$

Compressing the gas from sea level to 600 psi and increasing temperature from 70°F to 500°F reduces the relative humidity from 80% to *1.7%*.

### Dew Point of High Pressure Air

The dew point of the compressed air is determined by using the humidity ratio:

$P_{tot} = Total \ Pressure = 600 \ psi$

$$\omega = Humidity \ Ratio = 0.0125 \frac{lb \ vapor}{lb \ dry \ air}$$

From Table 6-4 for 60°C to 150°C:

$A = 7.96681$

$B = 1668.21$

$C = 228$

$$P_v = \frac{\omega \, P_{tot}}{0.622 + \omega} = \frac{0.0125 \times 600 psi}{0.622 + 0.0125} = 11.8 \, psi = 611 \, mmHg \quad \text{(6-33)}$$

$$T_{dew} = \frac{B}{A - \log P_v} - C = \frac{1668.21}{7.96681 - \log(611 mmHg)} - 228 \quad \text{(6-35)}$$

$$T_{dew} = 94°C = 201°F$$

The dew point of the high-pressure air is *201°F*. Since the system operating temperature of 120°F is less than this dew point, condensed moisture is likely in the pressure transducer cavity.

## 6.4.3   Humidity Life Equivalence

### Analysis

The relative life of components under humidity is a significant concern for non-hermetic parts such as plastic-encapsulated microcircuits (PEMs) since moisture on the surface of the die may result in die corrosion, electromigration, etc.. A method for relative humidity life was developed by Peck [57], which was correlated to the performance of epoxy microcircuit packages. This method uses temperature and relative humidity to determine the life ratio, as follows:

$$L \propto \phi^n \exp\left(\frac{E_a}{k \, T}\right) \quad \text{(6-36)}$$

$$\frac{L_1}{L_2} = \left(\frac{\phi_1}{\phi_2}\right)^n \exp\left(\frac{E_a}{k \, T_1} - \frac{E_a}{k \, T_2}\right) \quad \text{(6-37)}$$

$L, L_1, L_2 = Life \ at \ Level \ 1, 2$

$\phi, \phi_1, \phi_2 = Relative \ Humidity \ at \ Level \ 1, 2$

$E_a = Activation \ Energy = 0.77 \ eV \ (for \ epoxy \ microcircuits) \ [57]$

$n = Humidity \ Exponent = -3 \ (for \ epoxy \ microcircuits) [57]$

$k = Boltzmann \ Constant = 8.62 \times 10^{-5} \frac{eV}{°K} \ [52]$

$T, T_1, T_2 = Absolute \ Temperature \ at \ Level \ 1,2 \ (°K)$

It must be noted that the activation energy and humidity exponent are only given for epoxy microcircuits. Values for other materials and components as well as the validity of the correlation relationship should be determined for these items on a case-by-case basis.

**Underlying Assumptions**

- Humidity model described by Peck [57] is valid

### 6.4.4    Typical Humidity Life Equivalence

The humidity performance of plastic-encapsulated electronic components is to be determined in an accelerated life test. Determine test duration at 70°C and 90% relative humidity to simulate 50,000 hours of exposure at 25°C and 60% humidity.

**Analysis**

This method uses temperature and relative humidity to determine life equivalences:

$L_2 = Desired\ Life = 50,000\ hours$

$\phi_1 = Relative\ Humidity\ in\ Test\ Environment = 90\% = 0.9$

$\phi_2 = Relative\ Humidity\ in\ Desired\ Environment = 60\% = 0.6$

$T_1 = Temperature\ in\ Test\ Environment = 70°C = 343°K$

$T_2 = Temperature\ in\ Desired\ Environment = 25°C = 298°K$

$k = Boltzmann\ Constant = 8.62x10^{-5} \frac{eV}{°K}$ [52]

From Peck [57] for epoxy microcircuits:

$E_a = Activation\ Energy = 0.77\ eV$

$n = Humidity\ Exponent = -3$

$$\frac{L_1}{L_2} = \left(\frac{\phi_1}{\phi_2}\right)^n \exp\left(\frac{E_a}{k\ T_1} - \frac{E_a}{k\ T_2}\right)$$

$$\frac{L_1}{L_2} = = \left(\frac{0.9}{0.6}\right)^{-3} \exp\left(\frac{0.77eV}{8.62x10^{-5}\frac{eV}{°K}x343°K} - \frac{0.77eV}{8.62x10^{-5}\frac{eV}{°K}x298°K}\right) \quad (6\text{-}37)$$

$$\frac{L_1}{L_2} = \frac{L_1}{50,000\ hours} = 0.00580$$

$$L_1 = 0.00580x50,000\ hours = \textbf{290 hours}$$

A humidity test of *290 hours at 70°C* and 90% relative humidity will therefore simulate 50,000 hours at 25°C and 60% relative humidity.

# 6.5    PRESSURE-DROP ANALYSIS

Pressure drop is a concern in forced convection cooled electronics (see 3.6.2) where a cooling fluid is passed through a tube or plenum. The configuration needs to provide sufficient flow rate to meet cooling needs consistent with the available supply pressure. It should be mentioned that the methods described here do not include the effect of rapid changes in diameter, since the relationships are complex and there are several different approaches. Information on diameter contractions and expansions may be found in appropriate fluid mechanics literature [54], [59].

## 6.5.1    Pressure-Drop Relationship

The pressure drop [54] in a tube is given by the following relationship:

$$\Delta p = f \frac{l}{d} \rho \frac{V^2}{2} \qquad (6\text{-}38)$$

$$V = \frac{\dot{m}}{\rho A} \qquad (6\text{-}39)$$

$\Delta p$ = *Pressure Drop*
$f$ = *Friction Factor (see text and equation 6-40)*
$l$ = *Tube Length*
$d$ = *Tube Diameter*
$\rho$ = *Fluid Density*
$V$ = *Fluid Velocity*
$\dot{m}$ = *Mass Flow Rate*
$A$ = *Flow Cross-Sectional Area*

## 6.5.2    Friction Factor

The friction factor [58] used in equation 6-38 depends upon the pipe roughness and the Reynolds number. For flow in smooth tubes, the friction factor can be expressed as follows:

$$f = \frac{K}{\text{Re}^n} \tag{6-40}$$

$$\text{Re} = \frac{d\,V}{\nu} = \frac{\rho\,d\,V}{\mu} \tag{6-7}$$

$f$ = Friction Factor
Re = Reynolds Number
$K$, $n$ = Constants (see text and Table 6-5)
$d$ = Tube Diameter
$V$ = Fluid Velocity
$\nu$ = Kinematic Viscosity
$\rho$ = Fluid Density
$\mu$ = Absolute Viscosity

Typical values for the constants used in equation 6-40 for flow in a tube are listed in Table 6-5. The characteristic dimension for these relationships is the tube diameter.

## Table 6-5 - Equation 6-6 Constants for Friction in Tube [58]

| Re | K | n | Type of Flow |
|---|---|---|---|
| < 2,300 | 64 | 1 | Laminar |
| 2,300 – 20,000 | 0.316 | 0.25 | Turbulent |
| > 20,000 | 0.184 | 0.20 | Turbulent |

### 6.5.3    Effect of Bends in Cooling Path

Smooth bends in the cooling path increase the pressure drop over that expected for the same length of a straight tube [59]. This increased pressure drop may be determined for turbulent flow by determining a bend-loss factor (B) corresponding to the bend radius and angle:

$$\Delta p = B\,\rho\,\frac{V^2}{2} \tag{6-41}$$

$$B = \frac{\beta}{\text{Re}^{0.17}} \tag{6-42}$$

$\Delta p$ = *Pressure Drop*

$B$ = *Bend-Loss Factor*

$\rho$ = *Fluid Density*

$V$ = *Fluid Velocity*

$\beta$ = *Bend Factor (see Table 6-6)*

Re = *Reynolds Number (see equation 6-7)*

It must be noted that the bend-loss factor (B) includes the effect of the length of the tube within the bend, so only the straight portions of the tube need be included in length calculations. The above relationships are only valid for Reynolds numbers above the values shown in Table 6-6. Relationships for other Reynolds numbers and a more detailed treatment of tube bends may be found in Reference [59].

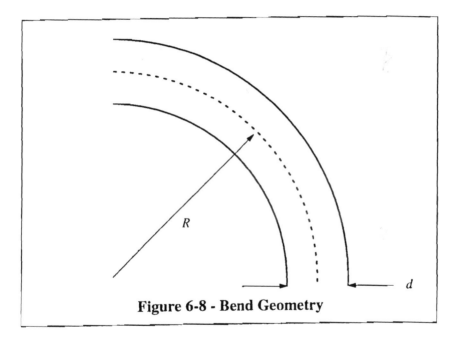

**Figure 6-8 - Bend Geometry**

## 6.5.4    Non-Circular Tubes

Approximate results for non-circular tubes may be approximated by using the hydraulic diameter in place of the tube diameter in the above equations. The hydraulic diameter is defined as follows:

$$d_h = \frac{4\,A}{P} \qquad\qquad (6\text{-}43)$$

$d_h$ = Hydraulic Diameter
$A$ = Tube Cross-Sectional Area
$P$ = Tube Perimeter

The hydraulic diameter is equal to the physical diameter for a circular cross-section. The approximation will be more accurate if the deviation from a circular cross-section is minor, although reasonably good results have been obtained for a rectangular cross-section [60]. More detailed treatment of rectangular ducts may be found in Reference [59].

| Table 6-6 - Equation 6-42 Bend Factors [59] | | | | |
|---|---|---|---|---|
| Bend Radius to Tube Diameter Ratio (R/d, see Figure 6-8) | Reynolds Number | Bend Factor (β) for Various Bend Angles | | |
| | | 45° | 90° | 180° |
| 1.8 | >4,000 | 0.64 | 0.96 | 1.10 |
| 2.5 | >4,000 | 0.49 | 0.71 | 0.91 |
| 3 | >4,000 | 0.43 | 0.62 | 0.88 |
| 4 | >5,760 | 0.36 | 0.54 | 0.88 |
| 5 | >9,000 | 0.33 | 0.51 | 0.89 |
| 6 | >12,960 | 0.31 | 0.49 | 0.90 |
| 8 | >23,040 | 0.29 | 0.47 | 0.92 |
| 10 | >36,000 | 0.28 | 0.47 | 0.94 |

## 6.5.5   Typical Pressure-Drop Analysis

For the smooth tube configuration shown in Figure 6-9, determine the pressure drop for flow of 70°F water at 1,000 lbm/hr.

**Fluid Velocity**

The fluid velocity is determined from the mass flow rate as follows:

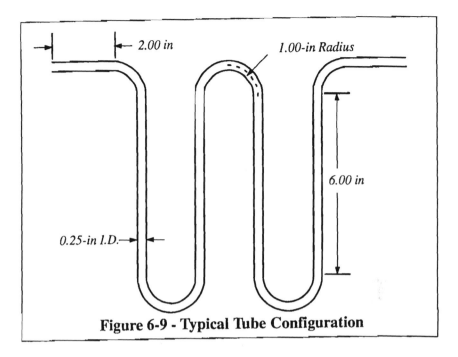

**Figure 6-9 - Typical Tube Configuration**

$\dot{m}$ = *Mass Flow Rate* = $1,000 \; \frac{lbm}{hr}$

$d$ = *Tube Diameter* = $0.25 \; in$

$A$ = *Flow Cross-Sectional Area* = $\pi \frac{d^2}{4} = \pi \frac{(0.25in)^2}{4} = 0.0491 \; in^2$

For water at 70°F [53]:

$\rho$ = *Fluid Density* = $62.27 \; \frac{lbm}{ft^3}$

$$V = \frac{\dot{m}}{\rho A} = \frac{1,000 \; \frac{lbm}{hr} x \frac{hr}{3,600sec}}{62.27 \; \frac{lbm}{ft^3} x 0.0491 \; in^2 x \frac{ft^2}{144in^2}} = 13.1 \frac{ft}{sec} \qquad (6\text{-}39)$$

**Reynolds Number**

The fluid velocity is then used to calculate the Reynolds number:

$V = Fluid\ Velocity = 13.1\frac{ft}{sec}$

$d = Tube\ Diameter = 0.25\ in$

For water at 70°F [53]:

$\rho = Fluid\ Density = 62.27\ \frac{lbm}{ft^3}$

$\mu = Absolute\ Viscosity = 2.37\ \frac{lbm}{ft\text{-}hr}$

$\mu = Kinematic\ Viscosity = \frac{\mu}{\rho} = 0.0381\ \frac{ft^2}{hr}$

$$\mathrm{Re} = \frac{d\ V}{\nu} = \frac{0.25in x\frac{ft}{12in}x13.1\frac{ft}{sec}x\frac{3.600sec}{hr}}{0.0381\ \frac{ft^2}{hr}} = 25,787 \qquad (6\text{-}7)$$

**Friction Factor**

The friction factor is determined from the Reynolds number:

$\mathrm{Re} = Reynolds\ Number = 25,787$

From Table 6-5 for Reynolds number of 25,787:

$K = 0.184$

$n = 0.2$

$$f = \frac{K}{\mathrm{Re}^n} = \frac{0.184}{25,787^{0.2}} = 0.0241 \qquad (6\text{-}40)$$

**Pressure Drop in Straight Portions of Tube**

The friction factor and fluid velocity is used to determine the pressure drop in the straight portions of the tube:

$f = Friction\ Factor = 0.0241$

$l = Total\ Tube\ Length\ (straight\ portions) = 4x6in + 2x2in = 28\ in$

$V = Fluid\ Velocity = 13.1\frac{ft}{sec}$

$d = Tube\ Diameter = 0.25\ in$

$\rho = Fluid\ Density = 62.27\ \frac{lbm}{ft^3}$

$$\Delta p = f\frac{l}{d}\rho\frac{V^2}{2} = 0.0241\frac{28in}{0.25in}62.27\frac{lbm}{ft^3}\frac{\left(13.1\frac{ft}{sec}\right)^2}{2}\frac{lbf}{32.2lbm\frac{ft}{sec^2}} = 447.9\frac{lbf}{ft^2}$$

(6-38)

$$\Delta p = 447.9\frac{lbf}{ft^2}\left(\frac{ft}{12in}\right)^2 = 3.11\,psi$$

### Pressure Drop in Curved Portions of Tube

The Reynolds number, and bend radius to tube diameter ratio are used to determine the pressure drop in the curved regions of the tube:

$$\rho = Fluid\ Density = 62.27\frac{lbm}{ft^3}$$

$$V = Fluid\ Velocity = 13.1\frac{ft}{sec}$$

$$Re = Reynolds\ Number = 25,787$$

$$\frac{R}{d} = Bend\ Radius\ to\ Tube\ Diameter\ Ratio = \frac{1in}{0.25in} = 4$$

### 90° bends

For a single 90° bend:

From Table 6-6 for R/d = 4 and 90° bend:

$$B = Bend\ Loss\ Factor = 0.54$$

$$B = \frac{\beta}{Re^{0.17}} = \frac{0.54}{25,787^{0.17}} = 0.0960$$

(6-42)

$$\Delta p = B\,\rho\frac{V^2}{2} = 0.0960x62.27\frac{lbm}{ft^3}\frac{\left(13.1\frac{ft}{sec}\right)^2}{2}\frac{lbf}{32.2lbm\frac{ft}{sec^2}} = 15.93\frac{lbf}{ft^2}$$

(6-41)

$$\Delta p = 15.93\frac{lbf}{ft^2}\left(\frac{ft}{12in}\right)^2 = 0.1106\,psi$$

180° bends

For a single 180° bend:

From Table 6-6 for R/d = 4 and 180° bend:

    $B = Bend\ Loss\ Factor = 0.88$

$$B = \frac{\beta}{Re^{0.17}} = \frac{0.88}{25,787^{0.17}} = 0.1565 \qquad (6\text{-}42)$$

$$\Delta p = B\,\rho\frac{V^2}{2} = 0.1565x62.27\frac{lbm}{ft^3}\frac{\left(13.1\frac{ft}{sec}\right)^2}{2}\frac{lbf}{32.2lbm\frac{ft}{sec^2}} = 25.97\frac{lbf}{ft^2}$$

$$\Delta p = 25.97\frac{lbf}{ft^2}\left(\frac{ft}{12in}\right)^2 = 0.1803\,psi \qquad (6\text{-}41)$$

**Total Pressure Drop**

The total pressure drop is the sum of the pressure drop in the straight portions, two 90° bends and three 180° bends:

    $\Delta p = 3.11psi + 2x0.1106psi + 3x0.1803psi = \mathbf{3.87\,psi}$

Therefore, the pressure drop in the given tube configuration (Figure 6-9) for 1,000 lbm/hr of water at 70°F is *3.87 psi*.

# 6.6    SIMILARITY CONSIDERATIONS

## 6.6.1    Background

In some cases, test data is available for one configuration but results are desired for another similar configuration. For example, suppose the pressure drop of a fuel-cooled electronic chassis is required to be below a certain value at a certain flow rate but it is not practical to conduct testing with fuel. The pressure drop for the fuel-cooled configuration may be determined by conducting testing with another fluid (such as water) and using similarity relationships to determine the pressure drop of the fuel. Since the friction factor depends upon the Reynolds number (see equation 6-40) and the loss factor for tube bends depends upon the geometry and Reynolds number (see equation 6-42), the flow will be similar if the Reynolds numbers are the same. Similarity equations can also be derived for other relationships by expressing the governing physical equations in non-dimensional form.

## 6.6.2    Fluid Flow Similarity

The pressure drop of a fluid at a given flow rate may be determined from testing of another fluid as follows:

1. Determine the Reynolds number at the given flow rate for the desired fluid

2. Determine the flow rate required for a test fluid to match the Reynolds number from step 1

3. Determine a system pressure drop constant based upon testing using the test fluid at the flow rate determined in step 2

4. Use the system pressure drop constant to determine the pressure drop of the desired fluid

**Reynolds Number of Desired Fluid**

The Reynolds number of the desired fluid is determined as follows:

$$\text{Re} = \frac{d\,V_d}{v_d} = \frac{\rho_d\,d\,V_d}{\mu_d} \qquad (6\text{-}7)$$

$$V_d = \frac{4\dot{Q}_d}{\pi d^2} \qquad (6\text{-}44)$$

Re $=$ *Reynolds Number*
$d =$ *Tube Diameter*
$V_d =$ *Velocity of Desired Fluid*
$v_d =$ *Kinematic Viscosity of Desired Fluid*
$\rho_d =$ *Fluid Density of Desired Fluid*
$\mu_d =$ *Absolute Viscosity of Desired Fluid*
$\dot{Q}_d =$ *Volume Flow Rate of Desired Fluid*

**Test Flow Rate**

The required flow rate of a test fluid for a pressure-drop evaluation is determined by matching the Reynolds number as follows:

$$\dot{Q}_t = V_t \frac{\pi d^2}{4} \qquad (6\text{-}45)$$

$$V_t = \frac{v_t\,\text{Re}}{d} = \frac{\mu_t\,\text{Re}}{\rho_t\,d} \qquad (6\text{-}46)$$

$\dot{Q}_t$ = *Volume Flow Rate of Test Fluid*

$V_t$ = *Velocity of Test Fluid*

$d$ = *Tube Diameter*

$v_t$ = *Kinematic Viscosity of Test Fluid*

Re = *Reynolds Number  (see equation 6-7 from previous section)*

$\rho_t$ = *Fluid Density of Test Fluid*

$\mu_t$ = *Absolute Viscosity of Test Fluid*

## Pressure-Drop Constant

Equations 6-38 and 6-41 may be combined and written in terms of a single constant (C) as follows:

$$\Delta p = C \, \rho \frac{V^2}{2} \qquad (6\text{-}47)$$

$\Delta p$ = *Pressure Drop*

$C$ = *Pressure-Drop Constant (function of Reynolds number and geometry)*

$\rho$ = *Fluid Density*

$V$ = *Fluid Velocity*

Solving for the pressure-drop constant and evaluating at the test flow rate and pressure drop:

$$C = \frac{2 \, \Delta p_t}{\rho_t \, V_t^2} \qquad (6\text{-}48)$$

$C$ = *Pressure-Drop Constant*

$\Delta p_t$ = *Tested Pressure Drop*

$\rho_t$ = *Fluid Density of Test Fluid*

$V_t$ = *Velocity of Test Fluid  (see equation 6-46)*

Since the pressure-drop constant depends upon the Reynolds number and geometry which are the same for the desired fluid and the test fluid, the pressure-drop constant will also be the same for both fluids.

## Pressure Drop of Desired Fluid

The pressure drop of the desired fluid is determined by evaluating equation 6-47 for the fluid at the appropriate flow rate:

$$\Delta p_d = C \, \rho_d \frac{V_d^2}{2} \tag{6-47}$$

$\Delta p_d$ = *Pressure Drop of Fluid at Chosen Flow Rate*
$C$ = *Pressure Drop Constant (function of Reynolds number and geometry)*
$\rho_d$ = *Fluid Density*
$V_d$ = *Fluid Velocity (see equation 6-44)*

## 6.6.3    Deriving Similarity Relationships

Although similarity relationships have traditionally been applied to fluid flow, as was just described, similarity can be applied to other situations [54]. One way of developing a similarity relationship is to express all of the variables in an equation as a product of a dimensionless variable and a dimensioned reference parameter. Simplifying the resulting equation by grouping the reference parameters together leads to the establishment of dimensionless groups. If this method is applied to equation 6-47, the following relationships are developed:

Using:
$$\Delta p = \Delta p_0 \, \Delta p^* \quad \text{(pressure drop)}$$
$$V = V_0 \, V^* \quad \text{(velocity)}$$

Substituting into equation 6-47:
$$\Delta p_0 \, \Delta p^* = C \, \rho \frac{(V_0 \, V^*)^2}{2} \tag{6-49}$$

Simplifying and collecting reference parameters:
$$\Delta p^* = C \left( \frac{\rho V_0^2}{\Delta p_0} \right) \frac{V^{*2}}{2} \tag{6-50}$$

If the reference parameters are chosen such that the dimensionless parameters are unity, and the resulting equation is solved for C:
$$C = 2 \left( \frac{\Delta p_0}{\rho V_0^2} \right) \tag{6-51}$$

$\Delta p_0$ = *Reference Pressure Drop*
$\Delta p^*$ = *Dimensionless Pressure Drop*
$\rho$ = *Density*
$V_0$ = *Reference Velocity*
$V^*$ = *Dimensionless Velocity*

It can be seen from the above that equation 6-51 is the same as equation 6-48.

The above approach can also be used with partial or ordinary differential equations to establish similarity relationships. This allows the relative performance of a complex system to be determined without actually solving the differential equations. Using the partial differential equation for conduction in a solid with constant properties:

$$k\left(\frac{\partial^2 T}{\partial x^2} + \frac{\partial^2 T}{\partial y^2} + \frac{\partial^2 T}{\partial z^2}\right) + \dot{q} = \rho c \frac{\partial T}{\partial \tau} \qquad (6\text{-}52)$$

$k$ = Thermal Conductivity

$T$ = Temperature

$x, y, z$ = Cartesian Coordinates

$\dot{q}$ = Heat Generation Rate per Unit Volume

$\rho$ = Density

$c$ = Specific Heat

$\tau$ = Time

Using dimensionless representations for the varying quantities:

Let:

$$x = L_0 x^* \qquad\qquad T = T_0 T^*$$
$$y = L_0 y^* \qquad\qquad \dot{q} = \dot{q}_0 \dot{q}^*$$
$$z = L_0 z^* \qquad\qquad \tau = \tau_0 \tau^*$$

Substituting into equation 6-52:

$$k\left(\frac{\partial^2(T_0 T^*)}{\partial(L_0 x^*)^2} + \frac{\partial^2(T_0 T^*)}{\partial(L_0 y^*)^2} + \frac{\partial^2(T_0 T^*)}{\partial(L_0 y^*)^2}\right) + \dot{q}_0 \dot{q}^* = \rho c \frac{\partial(T_0 T^*)}{\partial(\tau_0 \tau^*)} \qquad (6\text{-}53)$$

Simplifying and collecting reference parameters:

$$\left[\frac{k\tau_0}{\rho c L_0^2}\right]\left(\frac{\partial^2 T^*}{\partial x^{*2}} + \frac{\partial^2 T^*}{\partial y^{*2}} + \frac{\partial^2 T^*}{\partial y^{*2}}\right) + \left[\frac{\dot{q}_0 \tau_0}{\rho c T_0}\right]\dot{q}^* = \frac{\partial T^*}{\partial \tau^*} \qquad (6\text{-}54)$$

$k$ = Thermal Conductivity

$\tau_0$ = Reference Time

$\tau^*$ = Dimensionless Time

$\rho$ = Density

$c$ = Specific Heat

$L_0$ = Reference Dimension
$x^*, y^*, z^*$ = Dimensionless Cartesian Coordinates
$T_0$ = Reference Temperature
$T^*$ = Dimensionless Temperature
$\dot{q}_0$ = Reference Heat Generation Rate per Unit Volume
$\dot{q}^*$ = Dimensionless Heat Generation Rate per Unit Volume

Equation 6-54 will be identical for the dimensionless variables if the dimensionless groups in square brackets are the same, and will have the same solution if the dimensionless boundary conditions are the same. If two configurations have the same shape but are different sizes and have the same (dimensionless) boundary conditions, the dimensionless groups can be used to develop test conditions for one configuration relative to the other. This may have application in the thermal evaluation of an integrated circuit package or plated-through hole based upon a larger than actual size model.

## 6.6.4    Typical Similarity Analysis

A 4X size acrylic model of the ceramic (alumina) microcircuit package shown in Figure 6-10, produces a temperature rise of 25°C at the center of the die pad for a power input of 1 Watt. Determine the die-pad-to-case thermal resistance for the ceramic package. Constant heat flux is applied to the die pad and the base of the package is isothermal. Insulation is applied such that the heat transfer from the other surfaces is negligible.

**Similarity Relationship**

Using the similarity relationship for conduction in a solid derived in section 6.6.3:

For:

$$x = L_0 x^* \qquad\qquad T = T_0 T^*$$
$$y = L_0 y^* \qquad\qquad \dot{q} = \dot{q}_0 \dot{q}^*$$
$$z = L_0 z^* \qquad\qquad \tau = \tau_0 \tau^*$$

The partial differential equation for conduction becomes:

$$\left[\frac{k\tau_0}{\rho c L_0^2}\right]\left(\frac{\partial^2 T^*}{\partial x^{*2}} + \frac{\partial^2 T^*}{\partial y^{*2}} + \frac{\partial^2 T^*}{\partial y^{*2}}\right) + \left[\frac{\dot{q}_0 \tau_0}{\rho c T_0}\right]\dot{q}^* = \frac{\partial T^*}{\partial \tau^*} \qquad (6\text{-}54)$$

$k$ = Thermal Conductivity
$\tau_0$ = Reference Time
$\tau^*$ = Dimensionless Time

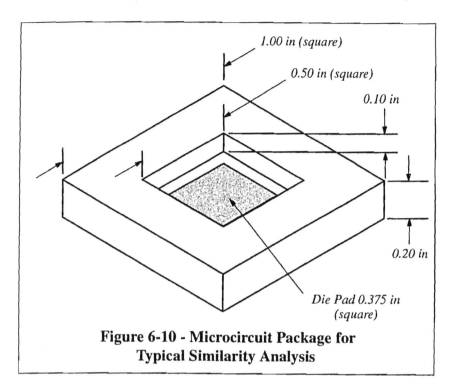

*1.00 in (square)*

*0.50 in (square)*

*0.10 in*

*0.20 in*

*Die Pad 0.375 in (square)*

**Figure 6-10 - Microcircuit Package for Typical Similarity Analysis**

$k$ = *Thermal Conductivity*

$\tau_0$ = *Reference Time*

$\tau^*$ = *Dimensionless Time*

$\rho$ = *Density*

$c$ = *Specific Heat*

$L_0$ = *Reference Dimension*

$x^*, y^*, z^*$ = *Dimensionless Cartesian Coordinates*

$T_0$ = *Reference Temperature*

$T^*$ = *Dimensionless Temperature*

$\dot{q}_0$ = *Reference Heat Generation Rate per Unit Volume*

$\dot{q}^*$ = *Dimensionless Heat Generation Rate per Unit Volume*

For steady-state conduction without internal heat generation equation 6-54 becomes:

$$\left[\frac{k\tau_0}{\rho c L_0^2}\right]\left(\frac{\partial^2 T^*}{\partial x^{*2}} + \frac{\partial^2 T^*}{\partial y^{*2}} + \frac{\partial^2 T^*}{\partial y^{*2}}\right) = 0 \qquad (6\text{-}55)$$

$$\frac{\partial^2 T^*}{\partial x^{*2}} + \frac{\partial^2 T^*}{\partial y^{*2}} + \frac{\partial^2 T^*}{\partial y^{*2}} = 0 \qquad (6\text{-}56)$$

So the dimensionless partial differential equation for steady-state conduction without internal heat generation is the same for all situations with similar shape. The solution of the above equation depends upon the boundary conditions.

## Boundary Conditions

### Base of Package

The base of package is isothermal and may be taken as the reference point so the dimensionless temperature boundary condition becomes the following:

$$T^* = 0 \quad (at\ base\ of\ package)$$

### Die-Pad Area

Using the relationship for heat flow in a solid (considering the Z-axis to be vertical):

Using:
$$a = L_0 a^*$$
$$z = L_0 z^*$$
$$T = T_0 T^*$$

The heat flow (power) at the interface is given as follows:

$$P = kA\frac{\partial T}{\partial z} \qquad (6\text{-}57)$$

Substituting, simplifying, and solving for the dimensionless boundary condition:

$$\frac{\partial T^*}{\partial z^*} = \frac{PL_0}{kAT_0} = \frac{PL_0}{k(L_0 a^*)^2 T_0} = \left[\frac{P}{kL_0 T_0}\right]\frac{1}{a^{*2}} \qquad (6\text{-}58)$$

$P = Heat\ Flow\ Rate\ (power)$

$k = Thermal\ Conductivity$

$A = Die\text{-}Pad\ Area = a^2$

$L_0 = Reference\ Dimension$

$a^*$ = *Dimensionless Die Pad Length/Width*

$z^*$ = *Dimensionless z*

$T_0$ = *Reference Temperature*

$T^*$ = *Dimensionless Temperature*

The boundary conditions in the die-pad area will be the same if the dimensionless group in the square brackets is the same and the shape of the die-pad area is the same. Since a scale model is being used, the second condition is satisfied.

### Other Areas

The boundary conditions in areas other than the die pad or base of package have no heat transfer so the temperature gradient normal to the surface is zero. This boundary condition is satisfied in both the evaluation model and the ceramic package.

### Evaluation

#### Scale Model Test Package

Using the length of the side of the package as the reference dimension and the temperature rise at the center of the package as the reference temperature, the dimensionless group for the 4X size acrylic test package is determined as follows:

$P$ = *Heat Flow Rate (power)* = 1 *W*

$T_0$ = *Reference Temperature* = 25° *C*

$L_0$ = *Reference Dimension* = 4 $x$ 1 *in* = 4 *in*

For acrylic [61]:

$$k = Thermal\ Conductivity = 1.44 \frac{Btu\text{-}in}{hr\text{-}ft^2\text{-}°F}$$

$$\left[ \frac{P}{kL_0T_0} \right] = \frac{1\ W}{1.44 \frac{Btu\text{-}in}{hr\text{-}ft^2\text{-}°F} x \left( \frac{ft}{12in} \right)^2 x 0.293 \frac{W\text{-}hr}{Btu} x 1.8 \frac{°F}{°C} x 4 in x 25°C} = 1.896$$

#### Ceramic Package

Using the same value for the bracketed dimensionless group but using the appropriate reference dimension and conductivity, the reference temperature for the ceramic package is determined as follows:

$P$ = *Heat Flow Rate (power)* = 1 *W*

$T_0$ = *Reference Temperature* = 25°*C*

$L_0$ = *Reference Dimension* = 1 *in*

For alumina ceramic [61]:

$$k = Thermal\ Conductivity = 14.5 \frac{Btu}{hr\text{-}ft\text{-}°F}$$

$$\left[\frac{P}{kL_0T_0}\right] = 1.896$$

Solving for $T_0$:

$$T_0 = \frac{P}{1.896\ kL_0}$$

$$T_0 = \frac{1\ W}{1.896 x 14.5 \frac{Btu}{hr\text{-}ft\text{-}°F} x \frac{ft}{12in} x 0.293 \frac{W\text{-}hr}{Btu} x 1.8 \frac{°F}{°C} x 1 in} = 0.828°C$$

$$\theta = \frac{T_0}{P} = \frac{0.828°C}{1\ W} = \mathbf{0.828 \frac{°C}{W}}$$

So the thermal resistance from the die pad to the base of the ceramic package is *0.828° C/W.*

## 6.7    ENERGY-BASED METHODS

### 6.7.1    Background

Energy-based methods use conservation of energy to establish the appropriate analytical relationships. Typically the initial energy state is determined and this is used to determine the distribution of energy in other conditions. An energy-balance method was used to determine the life of leaded chip-carriers under thermal cycling (see 5.2.3) and is also used in the derivation of compressible ideal-gas flow relationships (see 6.3.2). In developing any energy-based method, the following questions need to be answered:

- Where is the initial energy stored?
- How is the energy dissipated and/or transformed?
- Which energy state represents the worst-case situation for the analysis?

Often in the analysis of electronic packaging, the desired result is the deflection or stress developed in the hardware. This deflection or stress is typically the result of internal mechanical energy storage in, or accelerations applied to the hardware. The initial energy state corresponding to this internal energy storage may result from potential or kinetic energy. Typically it is assumed that all of the energy stored in resulting state is due to elastic internal storage which would represent a worst-case situation. Any motion of the body in the final state would result in some energy being kinetic which would reduce the elastic energy, resulting in less deflection/stress. Energy balance for a dropped box is illustrated in Figure 6-11 (details of the equations will be described later).

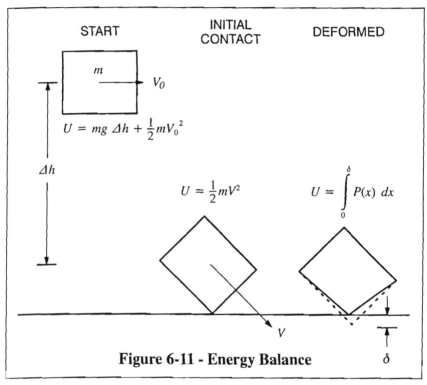

**Figure 6-11 - Energy Balance**

## 6.7.2    Drop-Test Analysis

Some electronic hardware is required to sustain a drop test. In these situations, the hardware is dropped from a given height onto a specified surface. Results of this situation may include the following:

- Damage to the external chassis due to excessive force (yielding and/or overstressing)

- Damage to internal equipment due to the resulting deceleration

## Elastic Energy from Potential Energy

Equivalence between elastic and potential energy would typically be used to determine the stresses and deflections resulting from the dropping of an object and can be expressed by the following relationships:

$$E_P = mg\, \Delta h \tag{6-59}$$

$$E_E = \frac{1}{2}K\delta^2 = \frac{1}{2}\frac{F^2}{K} \tag{6-60}$$

$$F = k\delta \tag{6-61}$$

Equating Potential and Elastic energy and solving for deflection/force:

$$\delta = \sqrt{\frac{m}{K}}\ \sqrt{2g\,\Delta h} \tag{6-62}$$

$$F = \sqrt{mK}\ \sqrt{2g\,\Delta h} \tag{6-63}$$

$$a = \frac{F}{m} = \sqrt{\frac{K}{m}}\ \sqrt{2g\,\Delta h} \tag{6-64}$$

$E_P$ = Potential Energy
$m$ = Mass (see text)
$g$ = Acceleration Due to Gravity
$\Delta h$ = Change in Height
$E_E$ = Elastic Energy
$K$ = Effective Spring Rate of Elastic Energy Storage
$\delta$ = Maximum Deflection
$F$ = Maximum Force
$a$ = Maximum Acceleration

Determination of the mass in the above equations is represented by the amount of mass supported by the elastic force in the elastic energy state. The effective spring rate of elastic energy storage may be determined by a classical or finite-element analysis in the region of impact (see 6.7.5). It is important to note that the above relationships are based upon elastic performance of the effective spring, calculations for plastic situations (such as permanent deformation of the chassis and/or surface being impacted) are described in 6.7.4.

**Underlying Assumptions**

- All potential energy is dissipated as elastic energy
- The system is at rest in the elastic energy state
- Impact between the bodies is quasi-static (dynamic effects neglected)

### 6.7.3    Impact Analysis of Equipment with Initial Velocity

**Elastic Energy from Potential and Kinetic Energy**

Equivalence between elastic and combined potential and kinetic energy would typically be used to determine the stresses and deflections resulting from impact from a falling object with an initial velocity. The energy balance in this situation can be expressed by the following relationships:

$$E_P = mg\,\Delta h \tag{6-59}$$

$$E_K = \frac{1}{2}mV^2 \tag{6-65}$$

$$U = E_P + E_K = mg\,\Delta h + \frac{1}{2}mV^2 \tag{6-66}$$

$$E_E = \frac{1}{2}K\delta^2 = \frac{1}{2}\frac{F^2}{K} \tag{6-60}$$

Equating total Kinetic and Potential, and Elastic energy and solving for deflection/force:

$$\delta = \sqrt{\frac{m}{K}}\,\sqrt{2g\,\Delta h + V^2} \tag{6-67}$$

$$F = \sqrt{mK}\,\sqrt{2g\,\Delta h + V^2} \tag{6-68}$$

$$a = \frac{F}{M} = \sqrt{\frac{K}{m}}\,\sqrt{2g\,\Delta h + V^2} \tag{6-69}$$

$E_P$ = *Potential Energy*

$m$ = *Mass (see text)*

$g$ = *Acceleration Due to Gravity*

$\Delta h$ = *Change in Height*

$E_K$ = *Kinetic Energy*

$V$ = *Velocity*

$U$ = *Total Initial Energy*

$E_E$ = *Elastic Energy*

$K$ = *Spring Rate of Elastic Energy Storage*
$\delta$ = *Maximum Deflection*
$F$ = *Maximum Force*
$a$ = *Maximum Acceleration*

As was the case with the potential energy equivalence, the mass in the above equations is represented by the amount of mass supported by the elastic force in the elastic energy state. As before, the effective spring rate may be determined by a classical or finite-element analysis in the region of impact (see 6.7.5). The above kinetic energy relationships also depend upon elastic performance of the effective spring. Calculations for plastic situations are described in 6.7.4.

### Underlying Assumptions

- All kinetic energy is dissipated as elastic energy
- The system is at rest in the elastic energy state
- Impact between the bodies is quasi-static (dynamic effects neglected)

## 6.7.4    Non-Elastic Energy Storage

### Background

In some situations, the drop or impact situation described above can result in yielding of the materials. In these cases additional deflection will be required to absorb the initial energy. The elastic energy in this case can be determined by integrating the force deflection curve:

$$U = E_M = \int_0^\delta P(x)\ dx \qquad (6\text{-}70)$$

$$F = P(\delta) \qquad (6\text{-}71)$$

$U$ = *Initial Energy (kinetic and/or potential)*
$E_M$ = *Mechanical Energy*
$\delta$ = *Deflection*
$P(x)$ = *Force-Deflection Function*
$x$ = *Integration Variable (deflection)*
$F$ = *Force*

The force-deflection relationship may be determined by a non-linear finite-element analysis. Since the integral in equation 6-70 is path-dependent, the non-linear analysis will typically require a series of force-deflection points to properly calculate the integral. Numerical integration would probably be necessary for results obtained from finite-element analysis.

**Impact of Corner of Chassis on Wood**

In many cases a drop test involves dropping an electronic chassis onto a wood surface. Typically a corner of the chassis contacts the wood leaving an indentation. The forces and accelerations resulting from this impact may be determined by expressing the crushed area as a function of deflection and calculating the energy absorbed as follows:

$$A(x) = C_A \, x^2 \tag{6-72}$$

$$P(x) = \sigma_c A(x) = \sigma_c C_A \, x^2 \tag{6-73}$$

$$U = E_M = \int_0^\delta P(x) \, dx = \int_0^\delta \sigma_c C_A \, x^2 dx \tag{6-74}$$

$$E_M = \frac{1}{3}\sigma_c C_A \, \delta^3 \tag{6-75}$$

$$\delta = \sqrt[3]{\frac{3U}{\sigma_c C_A}} \tag{6-76}$$

$$A = \sqrt[3]{\frac{9C_A U^2}{\sigma_c^2}} \tag{6-77}$$

$$F = \sqrt[3]{9C_A \sigma_c U^2} \tag{6-78}$$

$$a = \frac{F}{m} = \frac{\sqrt[3]{9C_A \sigma_c U^2}}{m} \tag{6-79}$$

$A(x) = $ *Indentation Area-Deflection Function*

$C_A = $ *Area-Deflection Constant (see text and Table 6-7)*

$x = $ *Integration Variable (deflection)*

$P(x) = $ *Force-Deflection Function*

$\sigma_c = $ *Compression Strength of Wood*

$U$ = *Initial Energy (kinetic and/or potential)*

$E_M$ = *Mechanical Energy*

$\delta$ = *Maximum Deflection*

$A$ = *Maximum Indentation Area*

$F$ = *Maximum Force*

$a$ = *Maximum Acceleration*

$m$ = *Equipment Mass*

The area-deflection constant ($C_A$) represents the proportionality between the indentation area and the deflection and is determined by geometry of the system. This constant depends upon the angle of the impact, but if the impact results from a straight vertical drop the impact corner must line up with the center of gravity. Based upon this condition, the area-deflection constant can be expressed in terms of the ratio of the center of gravity distances (see Figure 6-12). Values of the area-deflection constant for various center of gravity ratios are listed in Table 6-7.

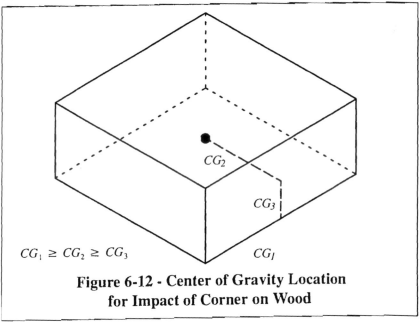

$$CG_1 \geq CG_2 \geq CG_3$$

**Figure 6-12 - Center of Gravity Location
for Impact of Corner on Wood**

Underlying Assumptions

- The wood has no elastic behavior

- The wood deflects as soon as the maximum compression stress is exceeded

- Only one corner is involved in the impact

- Deformation of the chassis is neglected
- All energy is dissipated in the wood
- The system is at rest when the wood is fully deflected
- Impact between the bodies is quasi-static (dynamic effects neglected)
- Impact corner aligns with center of gravity

| Table 6-7 - Area-Deflection Constants for Impact of Corner on Wood | | |
|:---:|:---:|:---:|
| **Center of Gravity Ratios (see Figure 6-12)** | | **Area-Deflection Constant ($C_A$)** |
| $CG_2/CG_1$ | $CG_3/CG_1$ | |
| 1.0 | 1.0 | 2.598 |
| 1.0 | 0.8 | 2.681 |
| 1.0 | 0.6 | 3.021 |
| 1.0 | 0.4 | 3.968 |
| 1.0 | 0.2 | 7.284 |
| 0.8 | 0.8 | 2.690 |
| 0.8 | 0.6 | 2.946 |
| 0.8 | 0.4 | 3.773 |
| 0.8 | 0.2 | 6.805 |
| 0.6 | 0.6 | 3.133 |
| 0.6 | 0.4 | 3.904 |
| 0.6 | 0.2 | 6.902 |
| 0.4 | 0.4 | 4.739 |
| 0.4 | 0.2 | 8.216 |
| 0.2 | 0.2 | 14.030 |

## 6.7.5    Stress Calculations

Once the forces and accelerations on an electronic enclosure have been determined, the corresponding stresses need to be calculated. These stresses can then
be used to determine if cracking, yielding, or other permanent damage have oc-

curred. In the case of elastic energy storage (see 6.7.2 and 6.7.3), the stresses in the area of impact can be determined from the finite-element model used to calculate the spring rate of the energy storage. Similarly, for non-elastic energy storage (see 6.7.4), the finite-element model used to determine the force-deflection relationship can be used to determine stress in the area of impact.

**Stresses due to Acceleration**

For stresses in areas of the electronic box away from the point of impact or for impact with wood, the resulting maximum acceleration can be used to determine the stresses (see 4.6.2). In many cases, the duration of the impact might be such that a mechanical shock analysis (see 4.5.5) is required to determine the internal response. The acceleration waveform may be obtained from the following differential equation:

$$\ddot{x} = \frac{d^2x}{dt^2} = -\frac{P(x)}{m} \qquad (6\text{-}80)$$

With the following initial conditions (time = 0):

$$\dot{x}_0 = \left(\frac{dx}{dt}\right)_0 = \sqrt{\frac{2U}{m}} \qquad\qquad x_0 = 0$$

$\ddot{x} = Acceleration$
$P(x) = Force\text{-}Deflection\ Function$
$m = Equipment\ Mass$
$\dot{x}_0, x_0 = Initial\ Velocity,\ Deflection$
$U = Initial\ Energy\ (kinetic\ and/or\ potential)$

If the force-deflection relationship is non-linear, the solution to the above differential equation may need to be determined numerically, but for elastic contact the equation may be simplified as follows:

$$\ddot{x} = \frac{d^2x}{dt^2} = -\frac{Kx}{m} \qquad (6\text{-}81)$$

This equation is the same as for an undamped spring-mass system (see 4.4.1) and has the following solution:

$$x = \frac{1}{\omega}\sqrt{\frac{2U}{m}}\sin(\omega t) \qquad (6\text{-}82)$$

$$\dot{x} = \sqrt{\frac{2U}{m}}\cos(\omega t) \qquad (6\text{-}83)$$

$$\ddot{x} = -\omega\sqrt{\frac{2U}{m}}\sin(\omega t) \qquad (6\text{-}84)$$

$$\omega = \sqrt{\frac{K}{m}} \qquad (6\text{-}85)$$

$\ddot{x}$ = *Acceleration*

$x$ = *Instantaneous Deflection*

$K$ = *Spring Rate of Elastic Energy Storage*

$m$ = *Equipment Mass*

$U$ = *Initial Energy (kinetic and/or potential)*

$\omega$ = *Circular Frequency*

Typical acceleration vs. time curves for elastic impact and impact with wood are shown in Figure 6-13. Although a shock analysis may be required to determine the response, the assumption of a quasi-static impact may still apply. The quasi-static assumption applies to the area of impact, whereas the need for shock analysis applies to the performance of internal hardware.

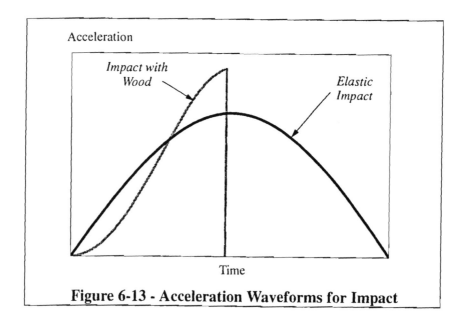

**Figure 6-13 - Acceleration Waveforms for Impact**

### Distribution of Stresses

For a given amount of energy, the maximum stress depends upon the distribution of energy within the body. For a bar under an axial impact, the quasi-static stress is uniform so the energy density within the bar is uniform. If a cantilevered beam is subjected to a similar transverse impact, the stress increases through the thickness and toward the root of the beam (see Figure 6-14). This non-uniform stress distribution results in a non-uniform energy distribution which increases the maximum stress corresponding to a given energy. So it can be expected for a given energy applied, the maximum stress is larger if a non-uniform stress distribution is present.

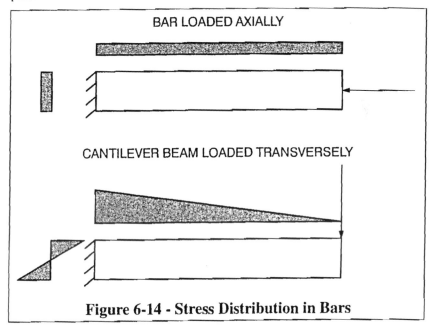

**Figure 6-14 - Stress Distribution in Bars**

## 6.7.6  Typical Drop-Test Analysis

A 10-in by 8-in by 6-in electronic chassis weighing 24 pounds is dropped onto an eastern white pine surface from a height of 6 feet with an initial downward velocity of 8 feet/sec. Determine the following:

- The maximum force on the corner of the chassis

- The maximum acceleration on the chassis

The center of gravity of the chassis is at the geometric center.

**Initial Energy**

The initial energy is the sum of the potential and kinetic energies as follows:

$\Delta h$ = Change in Height = 6 ft

$m$ = Mass = 24 lbm

$V$ = Initial Velocity = $8 \frac{ft}{sec}$

$g$ = Acceleration Due to Gravity = $32.2 \frac{ft}{sec^2}$

$$E_P = mg \, \Delta h = 24 lbm \frac{lbf}{32.2 lbm \frac{ft}{sec^2}} 32.2 \frac{ft}{sec^2} x 6 ft = 144 \, ft\text{-}lbf \qquad (6\text{-}59)$$

$$E_K = \frac{1}{2} mV^2 = \frac{1}{2} 24 lbm \frac{lbf}{32.2 lbm \frac{ft}{sec^2}} \left( 8 \frac{ft}{sec} \right)^2 = 23.8 \, ft\text{-}lbf \qquad (6\text{-}65)$$

$$U = E_P + E_K = mg \, \Delta h + \frac{1}{2} mV^2$$
$$\qquad (6\text{-}66)$$
$$U = 144 \, ft\text{-}lbf + 23.8 ft\text{-}lbf = 167.8 \, ft\text{-}lbf$$

**Chassis Force/Acceleration**

The chassis force and acceleration is determined by equating the total initial energy to that absorbed by the impact with the wood:

$U$ = Initial Energy (kinetic and/or potential) = $167.8 \, ft\text{-}lbf$

$m$ = Equipment Mass = 24 lbm

For eastern white pine [62]

$\sigma_c$ = Compression Strength of Wood = 440 psi

$CG_1$ = Largest CG Distance = 5 in

$CG_2$ = Intermeditate CG Distance = 4 in

$CG_3$ = Smallest CG Distance = 3 in

$\dfrac{CG_2}{CG_1} = \dfrac{4in}{5in} = 0.8$

$\dfrac{CG_3}{CG_1} = \dfrac{3in}{5in} = 0.6$

From Table 6-7 for the appropriate CG ratios, the following is determined:

$C_A$ = Area-Deflection Constant = 2.946

$$F = \sqrt[3]{9C_A\sigma_c U^2} = \sqrt[3]{9 \times 2.946 \times 440\,psi \times \left(\frac{12in}{ft}\right)^2 (167.8ft\text{-}lbf)^2}$$

$$F = 3,617\,lbf$$

(6-78)

$$a = \frac{F}{m} = \frac{3,617lbf}{24lbm}x\frac{32.2lbm\frac{ft}{sec^2}}{lbf} = 4,851\frac{ft}{sec^2}$$

(6-79)

$$a = 4,851\frac{\frac{ft}{sec^2}}{32.2\frac{ft}{sec}}\,2 = 151\,g$$

The maximum force on the corner of the chassis is *3,617 lbf* resulting in a maximum acceleration of *151 g*.

# 6.8   FACTORS IN RECENT DEVELOPMENTS

The methods described in this chapter are principally applicable to analysis at the chassis level and are not influenced by recent developments at the component level (see 1.1.3). Although humidity analysis, drop-test analysis, and similarity methods can include component-level factors, no special considerations are required to analyze these configurations.

# 6.9   VERIFICATION

Any results from analysis should be verified to help avoid any inaccuracies that might arise. This verification may be by testing, use of a simplified model, or other methods as appropriate. For information on verification, see Chapter 8.

# 6.10   OTHER ANALYSIS CHECKLIST

## 6.10.1   Applicable to All Analyses

☐ Has all source material (material properties, environmental data, etc.) been verified against the references?

## 6.10.2   Applicable to Fire-Resistance Analysis

☐ Has the calibration configuration been properly defined?

☐  Has the effect of temperature on material properties been considered?

### 6.10.3  Applicable to Pressure Transducer Rupture

☐  Has the effect of temperature on material properties been considered?

☐  Has the effect of restricting orifice on response time been considered?

☐  Has mass basis been used for the universal gas constant (R)?

### 6.10.4  Applicable to Humidity Analysis

☐  Are the units for pressure consistent with the vapor pressure equation (equation 6-31)?

☐  Are the Table 6-4 constants consistent with the temperature/dew point?

☐  Is the activation energy and humidity exponent for relative humidity life consistent with the materials under consideration?

### 6.10.5  Applicable to Pressure-Drop Analysis

☐  Has the effect of temperature on material properties been considered?

☐  Does the calculated tube length include only the straight sections of the tube?

### 6.10.6  Applicable to Similarity Analysis

☐  Do the similarity configurations have similar shape, or has the effect of shape been included in the similarity considerations?

☐  Have all appropriate system variables been expressed in a dimensionless form?

### 6.10.7  Applicable to Energy-Based Methods

☐  Have all appropriate initial energy states been identified?

☐  Have all the energy states for situation being analyzed been correctly identified?

☐  Has the system mass been properly considered?

☐  Has the effect of plastic deformation been considered (if applicable)?

## 6.10.8   Environmental Data Required

The following environmental data is typically required for various analyses. Typical units are provided for convenience, and care must be taken to ensure that the system of units is consistent with the material properties and model dimensions.

**Fire-Resistance Analysis**

☐   Heat input to calibration tube (W, Btu/hr, etc.)

☐   Temperature of gas (°C, °F, etc.)

☐   Surfaces of impingement of flame on chassis

**Pressure Transducer Rupture**

Heat Transfer due to Rupture

☐   Pressure of sensed gas at source (MPa, psi, etc.)

☐   Ambient pressure surrounding chassis (MPa, psi, etc.)

☐   Chassis temperature (°C, °F, etc.)

☐   Temperature of sensed gas at source (°C, °F, etc.)

Effect of Orifice on Response Time

☐   Temperature of gas in transducer (°C, °F, etc.)

**Humidity Analysis**

Humidity Relationships

☐   Pressure (mmHg, MPa, psi, etc.)

☐   Temperature (°C, °F, etc.)

☐   Relative humidity (dimensionless)

Relative Humidity Life

☐   Absolute temperature (°K, °R)

☐   Relative humidity (dimensionless)

☐   Duration of humidity (sec, min, hr, etc.)

## Pressure-Drop Analysis

One of the following:

☐   Fluid flow rate (kg/sec, lbm/hr, etc.)

☐   Fluid temperature (°C, °F, etc.)

☐   Fluid velocity (m/sec, ft/sec, etc.)

## Similarity Analysis

Environmental data required for similarity analysis depends upon the specifics of the similarity relationship. It must be reiterated that the units used in such analysis must be selected such that dimensionless groups are indeed dimensionless.

## Energy-Based Methods

### Generic Energy-Based Methods

☐   Initial energy (J, ft-lbf, etc.)

### Drop and Impact Analysis

☐   Initial velocity (m/sec, ft/sec, etc.)

☐   Change in height (m, cm, ft, in, etc.)

## 6.10.9   Material Properties Required

The following material properties are typically required for various analyses. Typical units are provided for convenience, and care must be taken to ensure that the system of units is consistent for any analysis. Although some material properties have been provided in this chapter, the reader is strongly encouraged to independently verify these values.

## Fire-Resistance Analysis

☐   Thermal conductivity (W/(m-°C), Btu/(hr-ft-°F), W/(in-°C), etc.)

☐   Density (kg/m$^3$, lbm/ft$^3$, etc.)

☐   Specific heat (Joule/(kg-°C), Btu/(lbm-°C), cal/(g-°C), etc.)

☐ Absolute viscosity of hot gas and (if applicable) cooling fluid (kg/(sec-m), lbm/(sec-in), etc.)

## Pressure Transducer Rupture

☐ Density (kg/m$^3$, lbm/ft$^3$, etc.)

☐ Specific heat (Joule/(kg-°C), Btu/(lbm-°C), cal/(g-°C), etc.)

☐ Specific heat ratio (dimensionless)

☐ Universal gas constant (Joule/(kg-°C), Btu/(lbm-°C), cal/(g-°C), etc., using mass basis)

## Humidity Analysis

### Humidity Relationships

Humidity relationships are based upon a mixture of air and water vapor so the material properties are included in the equations provided. If relationships for mixtures of other gases and vapors are desired, the appropriate molecular weights and vapor pressure relationships should be used accordingly.

### Relative Humidity Life

☐ Activation energy for performance of material in humidity (eV, etc.)

☐ Humidity exponent for material (dimensionless)

## Pressure-Drop Analysis

☐ Density (kg/m$^3$, lbm/ft$^3$, etc.)

☐ Absolute viscosity (kg/(sec-m), lbm/(sec-in), etc.)

## Similarity Analysis

Material properties required for similarity analysis depend upon the specifics of the similarity relationship. Once again, the units used in such analysis must be selected to maintain dimensionless groups as dimensionless.

## Energy-Based Methods

### Elastic Drop and Impact Analysis

☐ Acceleration due to gravity (m/sec$^2$, ft/sec$^2$, in/sec$^2$, etc.)

Impact of Chassis Corner on Wood

☐    Compression strength of wood (MPa, psi, etc.)

Additional Information

In addition to the properties listed above, the additional information is required if effective spring rates or non-linear force-deflection relationships need to be calculated (see 4.11.7).

☐    Modulus of elasticity (MPa, psi, etc.)

☐    Poisson's ratio (dimensionless)

☐    Modulus of plasticity (MPa, psi, etc.; if applicable)

☐    Yield stress (MPa, psi, etc.; if applicable)

## 6.11   REFERENCES

[50]    "Fire Testing of Flexible Hose, Tube Assemblies, Coils, Fittings, and Similar Components", AS1055 Revision C; SAE International, 400 Commonwealth Drive, Warrendale, PA; June 1994

[51]    Lichty, L.; Combustion Engine Processes; McGraw-Hill, 1967

[52]    Sonntag & Van Wylen; Introduction to Thermodynamics: Classical and Statistical; John Wiley & Sons, 1971

[53]    Holman, J.P.; Heat Transfer, Third Edition; McGraw-Hill, 1972

[54]    Sabersky, Acosta, Hauptmann; Fluid Flow, Second Edition; Macmillan, 1971

[55]    Kinsler & Frey; Fundamentals of Acoustics, Second Edition; John Wiley and Sons, 1963

[56]    Dean, J. (ed); Lange's Handbook of Chemistry, 11th Edition; McGraw-Hill, 1974

[57]    Peck, D.; "Comprehensive Model for Humidity Testing Correlation", IEEE/IEPS, 1986

[58]    Incropera & DeWitt; Fundamentals of Heat Transfer, John Wiley & Sons, 1981

[59]    General Electric; Fluid Flow Data Book; General Electric Corporate Research and Development, Schenectady, NY, 1982

[60]   Kays, W.M.; Convective Heat and Mass Transfer; McGraw-Hill, 1972

[61]   Materials Engineering, 1992 Materials Selector; Penton Publishing, Cleveland, OH; December 1991

[62]   Baumeister, et al. (ed.); Mark's Standard Handbook for Mechanical Engineers Eighth Edition; McGraw-Hill, 1978

[63]   Kays, W.M. & London, A.L.; Compact Heat Exchangers, Third Edition; McGraw-Hill, 1984

# Chapter 7
# Analysis of Test Data

## 7.1 BACKGROUND

Although a variety of analytical techniques have been described in previous chapters, many of these approaches include inherent assumptions and simplifications. In other cases, the environment and/or equipment is too complex to analyze within the available resources. In these situations, testing may be required to obtain the necessary data or verify the analytical results. Prior planning of test configurations should be conducted to obtain the maximum benefit and to facilitate data analysis. It should be noted that the intent of this chapter is not to describe testing to verify compliance to some minimum requirement, but to obtain test data that can be used in analytical models. This is especially apparent in the case of life testing, where failures are something desired for verification purposes and not consequences to be avoided.

## 7.2 PLANNING TESTS

Planning of a test includes the following:

- Overall test approach
- Required instrumentation
- Required test duration and level
- Number of samples required

Once the above information is determined, various test conditions may be modified to obtain the maximum benefit for the available resources.

### 7.2.1 Test Approach

Development of a test approach depends upon the specific performance area to be evaluated. If the thermal performance of an electronic configuration is to be evaluated, simulated heat loads may be applied to various areas and the temperatures measured. For vibration evaluations, the hardware may be placed on an electrodynamic vibration system and the resulting accelerations, strains, etc. can be measured. In other cases, the difference in temperature-induced strain between the surface of a PWB and a reference sample (see Figure 7-11) may be used to determine thermal expansion rates [64],[65]. Literature in the specific field of interest should be consulted for developing appropriate tests.

## 7.2.2   Instrumentation

### Background

Choice of instrumentation is an important part of planning a test. If the instrumentation is not chosen properly, the results may be misleading or even unusable. Depending upon the test desired, the instrumentation must have appropriate resolution, sensitivity, repeatability, and accuracy to provide satisfactory results. This performance information for the instrumentation is typically provided in the equipment specifications, but it is important to understand the significance of the instrumentation performance on test results.

### Resolution

Resolution refers to the ability of an instrument to discriminate between readings with close but similar values. With the advent of digital measurement systems, resolution is often expressed as a number of digits. A 5-digit instrument will have more resolution than a 3-digit instrument, and would be able to differentiate more readily between close but different signals. Some instruments provide the capability to increase or decrease resolution, usually by increasing or decreasing the sample time respectively. Resolution may be enhanced by performing a comparison between a reference sample and a device under test as illustrated by the strain-gage bridge (see Figure 7-1). If the reference gage and test gage are connected as shown and the output voltage ($V_{out}$) is set to zero (or "nulled") by adjusting $R_{adj}$, the output voltage will be given by the following relationship:

$$V_{out} = \Delta R_x \left( \frac{(V^+ - V^-)R_{ref}}{(R_{ref} + R_x + \Delta R_x)(R_{ref} + R_x)} \right) \qquad (7\text{-}1)$$

For small changes in resistance, equation 7-1 may be rewritten as follows:

$$V_{out} = \Delta R_x \left( \frac{(V^+ - V^-)R_{ref}}{(R_{ref} + R_x)^2} \right) \qquad (7\text{-}2)$$

$$\frac{\Delta R_x}{R_x} = (GF)\, \epsilon \qquad (7\text{-}3)$$

$$V_{out} = \left( \frac{(V^+ - V^-)R_{ref}R_x(GF)}{(R_{ref} + R_x)^2} \right) \epsilon \qquad (7\text{-}4)$$

$V_{out}$ = *Output Voltage*

$\Delta R_x$ = *Change in Resistance of Test Gage*

$V^+, V^-$ = *Positive, Negative Excitation Voltage*

$R_{ref}$ = *Nominal Resistance of Reference Gage*

$R_x$ = *Nominal Resistance of Test Gage*

$GF$ = *Gage Factor of Strain Gage*

$\epsilon$ = *Mechanical Strain*

Output voltage will be nearly proportional to the strain for small strain values. Although the bridge circuit allows the resolution to be enhanced by detecting small changes in resistance, the instrumentation must have sufficient sensitivity to detect the small signal that will be obtained. For typical 350-Ohm strain gages, an excitation of 10 Volts, and a gage factor of 2 the output voltage at 1% strain will be 0.050 Volts.

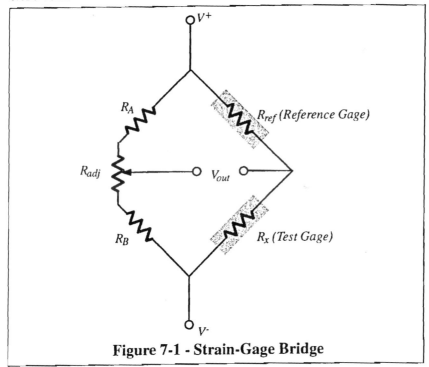

## Figure 7-1 - Strain-Gage Bridge

**Sensitivity**

Sensitivity is the ability to detect very small signals. If the sensitivity is not selected properly, the instrument may not respond to the variations expected in a typical test or may be overloaded by the signal. As an example, a volt-ohm meter (VOM) would not have sufficient sensitivity to measure a signal from a thermo-couple or the output of the strain-gage bridge described above.

### Repeatability

Repeatability refers to the capability to obtain the same reading for the same parameter after multiple trials. Repeatability reflects on the stability (lack of drift) within the instrumentation. Stability in the measurement system is important in situations such as transient thermal measurements of a chassis with a long time constant (see Figure 7-2), since as the system nears steady-state there is very little change in temperature over time. Drift can have an especially adverse effect on results if rate of change information is required (see Figure 7-2).

### Accuracy

Accuracy is the capability of an instrumentation system to produce measurements close to the actual standardized values. Typically the accuracy of an instrumentation system is maintained by periodic checks on calibration against a standard reference. The inaccuracies in the instrumentation system will be reflected as inaccuracies in the measured results. Ideally the accuracy should be commensurate with the resolution and repeatability of an instrumentation system. If a digital voltmeter had 5 digits of resolution but only had 1% accuracy, measurements made to the full 5 digits of resolution may be useful but could be misleading. This situation would be especially problematic if instrumentation was changed during a test.

### Other Considerations

Other considerations in the selection of test instrumentation include the input impedance, the use of AC vs. DC coupling, frequency response, etc. Other literature such as that supplied with the instrumentation or related engineering textbooks should be consulted for additional information on other items that may affect test instrumentation.

## 7.2.3    Test Levels and Duration

Selection of a test level is important for both performance and life testing.

### Performance Testing

Test levels in performance should represent actual usage levels as closely as possible, although higher test levels will usually allow results to be more readily measured. For example if a heat sink had a thermal resistance of 2°C/W and testing

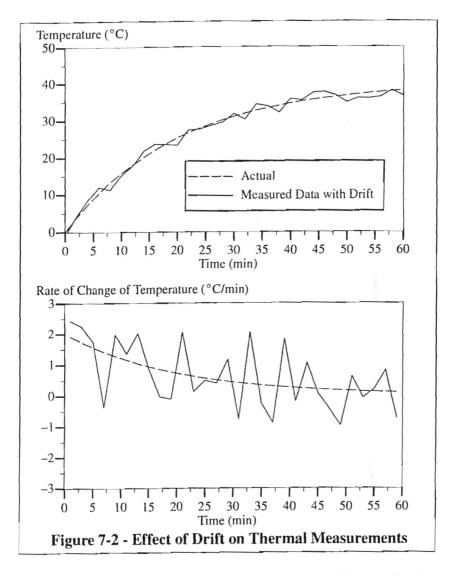

**Figure 7-2 - Effect of Drift on Thermal Measurements**

was conducted at 1 Watt, we could expect a temperature rise of 2°C. But if testing was conducted at 10 Watts, a 20°C temperature rise would result. Limitations on the use of higher test levels include the following:

- Non-linearities in the system performance

- Material limitations

## Non-Linearities

Non-linearities can arise in situations where the performance of a system deviates from the proportional performance observed in most systems. One situation in which non-linearities arise is free convection (see 3.6.2), where the convection coefficient increases with increasing temperature difference between the electronic chassis and ambient. Another case where non-linearities can be significant is in vibration (particularly with polymeric materials) where the damping increases at greater vibration levels. In both of these examples, the non-linearities may lead to over-prediction of performance at lower test levels if results are based upon testing at a higher level.

## Material Limitations

Material limitations include the thermal and structural limits of the component parts. In thermal testing, the temperatures should not exceed the temperature capabilities of the materials used in the system. For structural/vibration testing, the use of higher test levels should not result in yielding or failure of component parts.

## Life Testing

In life testing, increasing test levels usually tend to reduce the test duration required to demonstrate life (see Chapter 5). An accelerated test of 2,000 thermal cycles may simulate 50,000 hours of typical mission cycling if the test levels are properly selected. The ratio between the mission environment and the accelerated test environment is often referred to as the "acceleration factor". Limitations on accelerated tests include the following:

- Material limitations (described above)
- Non-linearities in the system performance
- Deviations in the fatigue curve

## Non-Linearities

The non-linearities in vibration performance described above can affect vibration life testing since an increased test level results in increased damping. Increased damping reduces the response vibration from that predicted by a linear model so the fatigue damage is also reduced. Vibration non-linearities can also affect resonant frequency which can result in a very significant effect if two resonant frequencies are very close to each other.

## Deviation in Fatigue Curve

In some cases, the fatigue curve can deviate from the typical power law used for fatigue life prediction. An example of this is the low-cycle fatigue of solder where the damage-to-fail changes with the cycles applied (see Figure 5-8). In this situation, life testing below 2,500 cycles (where damage-to-fail is increasing) may lead to overprediction of life above 2,500 cycles (where damage-to-fail is constant) unless the deviation in fatigue is accounted for (see low-cycle relative life of solder in 5.5.2). Deviation in the fatigue curve may also be noted in situations where the fatigue mechanism changes, such as variations in low-cycle fatigue due to creep and yielding.

## 7.2.4    Sample Size

Selection of sample size generally depends upon the expected variation in the measured parameter between test samples. If measurements are being made on a system that is readily predictable and material property variations are minimal, sample sizes may be small. An example of this may be thermal performance measurements of a metal heat sink in which good agreement is reached between the analysis and test. If discrepancies arise between analysis and test, additional test samples may be required.

In other situations, such as life testing, there is considerable variation between samples. In these situations, the sample size should be increased to allow the reliability distribution to be mapped (see 7.3.5). One way of verifying the sample size is to determine the probability of achieving a given number of failures in number of trials (samples) as follows [66]:

$$P(M) = \frac{N!}{M!(N-M)!} p^m (1-p)^{N-M} \qquad (7\text{-}5)$$

$P(M)$ = Probability of M Failures
$N$ = Sample Size
$M$ = Number of Failures
$p$ = Probability of Failure

If a test has a 5% probability of failure and there are only 10 samples, equation 7-5 will show that there is a 60% chance that no failures will be detected. Therefore, the sample size or the test duration should be increased so that the probability of failure increases. If the sample size is increased to 100, the probability of various numbers of failures is shown in Table 7-1. Increasing sample size to 100 results in only a 0.6% chance of no failures, compared to the 60% chance of no failures with a sample size of 10.

| Table 7-1 - Probability of M Failures for 100 Samples with 5% Failure Probability ||
| Number of Failures (M) | Probability of M Failures (%) |
| --- | --- |
| 0 | 0.59205 |
| 1 | 3.11607 |
| 2 | 8.11818 |
| 3 | 13.95757 |
| 4 | 17.81426 |
| 5 | 18.00178 |
| 6 | 15.00149 |
| 7 | 10.60255 |
| 8 | 6.48709 |
| 9 | 3.49013 |
| 10 | 1.67159 |
| 11 | 0.71982 |

### 7.2.5    Trade-offs

The effects of test level and duration, and sample size represent one of the major issues in test planning. The relative merits of increasing stress, test duration, and sample size must be weighed against schedule and cost objectives to obtain useful data. Selection of instrumentation used in testing is also significant in evaluating the relative merits. Although trade-offs may be made to achieve cost or schedule needs, it is important that the considerations described in 7.2.2 through 7.2.4 must be remembered to help avoid "useless" data.

## 7.3    DATA ANALYSIS

### 7.3.1    Thermal Performance Data Analysis

Thermal data analysis is usually conducted by applying a simulated or actual heat load and measuring the temperatures developed.

**Component (First) Level**

<u>Background</u>

Thermal measurements at the component level usually concentrate on obtaining the thermal resistance from the die pad or die surface to the outside of the package or heat sink. This is typically done by applying a simulated heat load in the die pad area and measuring the temperature at the die pad and heat sink (see Figure 7-3). Heat may be applied by a miniature blanket heater and temperature measured with a thermocouple, thermistor or similar device. Simulated thermal die are also available that provide a simulated heat input as well as incorporating thermal measurement capability. If size constraints present a problem, thermal measurements may be made on a scale model (see 6.6.4). If required to simulate actual operating conditions, the top surface of the component may be insulated. It should be noted that heat must be removed from the heat sink to avoid a steady increase in temperature due to the application of power. Measurements should be conducted at a number of power levels to provide verification.

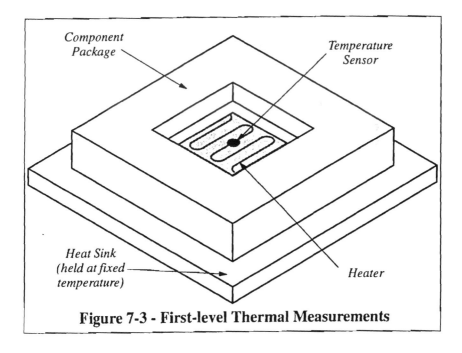

**Figure 7-3 - First-level Thermal Measurements**

Single-Point Analysis

Once results are obtained, the thermal resistance is determined as follows:

$$\theta = \frac{T_p - T_{hs}}{P} \qquad (7\text{-}6)$$

$\theta$ = *Thermal Resistance*
$T_p$ = *Temperature of Die Pad*
$T_{hs}$ = *Temperature of Heat Sink*
$P$ = *Power Applied*

If there is very little variation in the calculated thermal resistance at different power levels, the thermal resistance may be averaged to determine the overall resistance for the component. A more rigorous determination of thermal resistance may be obtained by the use of a linear regression.

Linear Regression Analysis

A linear regression is based upon minimizing the sum of the square of the deviation of each data point from an ideal linear relationship. This linear relationship may be based upon an intercept at the origin (zero temperature rise at zero power) as follows:

$$\theta = \frac{\Sigma(P_i \Delta T_i)}{\Sigma\left(P_i^2\right)} \qquad (7\text{-}7)$$

$$\Delta T_i = T_{p_i} - T_{hs_i} \qquad (7\text{-}8)$$

$\theta$ = *Thermal Resistance*
$P_i$ = *Power Applied at Level i*
$\Delta T_i$ = *Temperature Rise for Level i*
$T_{p_i}$ = *Temperature of Die Pad for Level i*
$T_{hs_i}$ = *Temperature of Heat Sink for Level i*

If there is a fixed temperature difference in the thermal sensors such that the temperature rise is not zero at zero power, the linear regression relationships are as follows [67]:

$$\theta = \frac{n\Sigma(P_i\Delta T_i) - \Sigma(P_i)\Sigma(\Delta T_i)}{n\Sigma\left(P_i^2\right) - (\Sigma(P_i))^2} \tag{7-9}$$

$$\Delta T_i = T_{P_i} - T_{hs_i} \tag{7-10}$$

$$T_{offs} = \frac{\Sigma(\Delta T_i) - \theta\Sigma(P_i)}{n} \tag{7-11}$$

$\theta$ = *Thermal Resistance*

$n$ = *Number of Test Levels (data points)*

$P_i$ = *Power Applied at Level i*

$\Delta T_i$ = *Temperature Rise for Level i*

$T_{P_i}$ = *Temperature of Die Pad for Level i*

$T_{hs_i}$ = *Temperature of Heat Sink for Level i*

$T_{offs}$ = *Temperature Offset of Thermal Sensors at Zero Power*

## PWB/Module (Second) Level

### Background

Thermal performance measurements at the module level typically involves temperature measurements of an actual module for verification of thermal analysis, or heat sink performance measurements using a simulated heat load. Usually when an actual module is tested, the resulting temperatures are directly compared against the thermal analysis or system requirements and no specific data analysis is required. Because of the complexity in the thermal analysis of through-hole technology heat sinks (see 3.5.2) and the need to have simulated heat loads applied at each potential component location, these configurations are also typically analyzed by the direct comparison method. A surface-mount technology heat sink is well suited to testing/analysis because of the relative simplicity of the thermal performance prediction (see 3.5.2).

### SMT Heat Sink Thermal Measurement

A surface-mount technology (SMT) heat sink (see Figure 3-10) takes heat from the components applied to the surface of the heat sink and conducts the heat along its length to the chassis. Testing of the SMT heat sink is accomplished by applying a simulated heat load on one face of the heat sink and measuring temperatures on the other face (see Figure 7-4). Similar to the component thermal measurements, heat must be removed from the heat sink edges to avoid a steady increase in temperature due to power. Measurements should also be conducted at a number of power levels to provide verification.

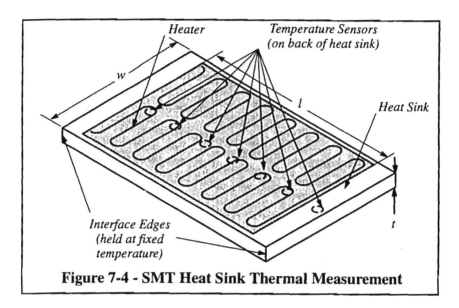

*Heater*  *Temperature Sensors (on back of heat sink)*

*w*

*l*

*Heat Sink*

*Interface Edges (held at fixed temperature)*

*t*

**Figure 7-4 - SMT Heat Sink Thermal Measurement**

<u>SMT Heat Sink Data Analysis</u>

A regression relationship can be developed for the SMT heat sink measurements by using the equation for theoretical performance of the heat sink:

$$\Delta T = P\frac{(l^2 - 4x^2)}{8k_{hs}lwt} \tag{3-10}$$

The deviation of an individual temperature data point is given as:

$$\delta_i = T_i - P_i\frac{(l^2 - 4x_i^2)}{8k_{hs}lwt} \tag{7-12}$$

The total (sum squared) deviation is given as:

$$\Sigma\delta_i^2 = \Sigma\left(T_i - P_i\frac{(l^2 - 4x_i^2)}{8k_{hs}lwt}\right)^2 \tag{7-13}$$

Letting $a = \dfrac{1}{8k_{hs}lwt}$ and expanding:

$$\Sigma\delta_i^2 = \Sigma\left(T_i - aP_i(l^2 - 4x_i^2)\right)^2 \tag{7-14}$$

$$\Sigma\delta_i^2 = \Sigma\left((T_i)^2 - 2al^2T_i\,P_i + 8aT_i\,P_ix_i^2 + a^2l^4P_i^2 - 8a^2l^2P_i^2x_i^2 + 16a^2P_i^2x_i^4\right)$$

Expressing as individual sums:

$$\Sigma\delta_i^2 = \Sigma(T_i)^2 - 2al^2\Sigma(T_i \ P_i) + 8a\Sigma(T_i \ P_ix_i^2) + a^2l^4\Sigma(P_i^2) \tag{7-15}$$
$$- 8a^2l^2\Sigma(P_i^2x_i^2) + 16a^2\Sigma(P_i^2x_i^4)$$

Differentiating on $a$ and setting equal to zero to find minimum sum squared deviation:

$$\frac{d(\Sigma\delta_i^2)}{da} = 0 \tag{7-16}$$

$$-2l^2\Sigma(T_i \ P_i) + 8\Sigma(T_i \ P_ix_i^2) + 2al^4\Sigma(P_i^2) - 16al^2\Sigma(P_i^2x_i^2) + 32a\Sigma(P_i^2x_i^4) = 0$$

Solving for $a$:

$$a = \frac{l^2\Sigma(T_i \ P_i) - 4\Sigma(T_i \ P_ix_i^2)}{l^4\Sigma(P_i^2) - 8l^2\Sigma(P_i^2x_i^2) + 16\Sigma(P_i^2x_i^4)} \tag{7-17}$$

The thermal conductivity may be determined as follows:

$$k_{hs} = \frac{l^4\Sigma(P_i^2) - 8l^2\Sigma(P_i^2x_i^2) + 16\Sigma(P_i^2x_i^4)}{8lwt(l^2\Sigma(T_i \ P_i) - 4\Sigma(T_i \ P_ix_i^2))} \tag{7-18}$$

$\Delta T$ = *Heat Sink Temperature Rise above Edge at Distance x from Center*
$x, x_i$ = *Distance from Center*
$P$ = *Total Module Power*
$l$ = *Heated Length/Span of Heat Sink*
$k_{hs}$ = *Thermal Conductivity of Heat Sink*
$w$ = *Width of Heat Sink*
$t$ = *Thickness of Heat Sink*
$\delta_i$ = *Temperature Deviation of Data Point i*
$T_i$ = *Measured Temperature Rise above Edge at Distance $x_i$ from Center* *
$P_i$ = *Total Module Power for Data Point i*

* - $\Delta$ is intentionally omitted for clarity in the resulting equations.

It must be noted that the above relationships only apply for temperature measurements within the heated length of the heat sink. The edge temperatures will generally be taken as the outer temperature sensors which are equidistant from the center. If these edge temperatures are not equal, a regression relationship may be

derived that also includes a linear temperature change along with the parabolic
relationship described above.

**Chassis (Third) Level**

<u>Background</u>

Thermal performance measurements at the chassis level typically involve mounting actual or simulated heat load modules in a chassis and measuring the thermal
performance (see Figure 7-5). Thermal sensor locations are determined by the
specific requirements of the thermal situation and may include the module rails,
key locations on the chassis wall, or other appropriate locations.

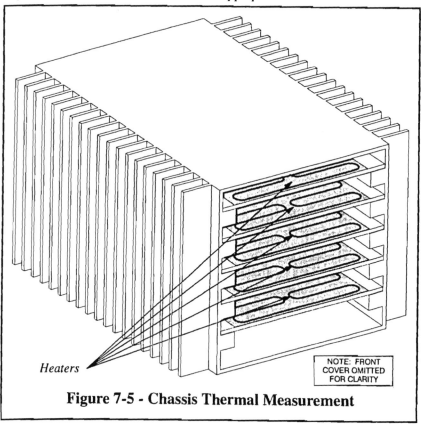

*Heaters*

NOTE: FRONT
COVER OMITTED
FOR CLARITY

**Figure 7-5 - Chassis Thermal Measurement**

<u>Chassis Thermal Measurement</u>

Usually when a chassis is tested, the resulting temperatures are directly compared
against the thermal analysis or system requirements and no specific data analysis

is required. Although component and module level thermal performance was principally dominated by conduction mechanisms, the chassis performance includes convection and radiation effects which are non-linear in temperature. It should be noted that there is a large class of electronic packaging configurations that use convection cooling of components and modules (see 3.4.1) which typically require experimentation to determine performance. Because of these non-linear effects, chassis thermal measurements should be conducted under conditions that closely match the actual thermal situation. The simulated heat loads should match the actual power dissipation as closely as possible and the ambient temperature should match the actual ambient. In matching the actual ambient temperature, care should be taken to ensure that the velocity of the surrounding air (or other fluid) is consistent with the desired situation. A chassis designed for free convection cooling will typically show much better thermal performance in a thermal chamber, at the same temperature, due to the action of the circulator fans included in most thermal chambers. Matching the radiation environment is also an important consideration. A chassis mounted in a free air environment will show better thermal performance when radiating to a open room than when surrounded by other electronic chassis operating at high temperature. It also should be noted that orientation of the test chassis is often important since free-convection coefficients are different for vertical and horizontal surfaces (see 3.6.2 and Table 3-2). Chassis thermal measurements may also include both transient and steady-state thermal measurements so it may also be important to match the thermal mass of the actual system. Since thermal measurement of chassis assemblies is conducted under conditions closely matching the actual operating environment, the analysis of the test data is usually limited to a direct comparison of temperatures.

## 7.3.2    Typical Thermal Data Analysis

A SMT heat sink 5-in wide by 0.125-in thick and 9-in long is heated for 8 in over the center by a blanket heater. Thermocouples are bonded on the opposite face from the blanket heater at the center and at 2-in and 4-in from each side of the center (see Figure 7-6). Power is applied from zero to 40 Watts in 10 Watt increments producing the data shown in Table 7-2. Determine the thermal conductivity of the heat sink material.

### Temperature Rise Above Edge

The heat sink temperature rise above the edge is determined by averaging the edge temperatures and subtracting from the temperature measurements at each power level (see Table 7-3).

| Table 7-2 - Typical Thermal Data | | |
|---|---|---|
| Power (W) | Distance from Center (in) | Temperature (°C) |
| 0 | -4 | 24.8 |
| 0 | -2 | 24.7 |
| 0 | 0 | 24.8 |
| 0 | 2 | 24.6 |
| 0 | 4 | 24.7 |
| 10 | -4 | 26.2 |
| 10 | -2 | 29.1 |
| 10 | 0 | 29.7 |
| 10 | 2 | 28.2 |
| 10 | 4 | 26.0 |
| 20 | -4 | 26.7 |
| 20 | -2 | 31.9 |
| 20 | 0 | 34.0 |
| 20 | 2 | 32.1 |
| 20 | 4 | 27.4 |
| 30 | -4 | 27.7 |
| 30 | -2 | 35.7 |
| 30 | 0 | 38.6 |
| 30 | 2 | 35.6 |
| 30 | 4 | 27.5 |
| 40 | -4 | 28.6 |
| 40 | -2 | 40.0 |
| 40 | 0 | 43.0 |
| 40 | 2 | 39.3 |
| 40 | 4 | 29.0 |

| Table 7-3 - Temperature Rise Calculation | | | | |
|---|---|---|---|---|
| Power (P$_i$, W) | Distance from Center (x$_i$, in) | Temperature (°C) | Average Edge Temperature (°C) | Temperature Rise (T$_i$, °C) |
| 0 | -4 | 24.8 |  | 0.05 |
| 0 | -2 | 24.7 |  | -0.05 |
| 0 | 0 | 24.8 | 24.75 | 0.05 |
| 0 | 2 | 24.6 |  | -0.15 |
| 0 | 4 | 24.7 |  | -0.05 |
| 10 | -4 | 26.2 |  | 0.10 |
| 10 | -2 | 29.1 |  | 3.00 |
| 10 | 0 | 29.7 | 26.10 | 3.60 |
| 10 | 2 | 28.2 |  | 2.10 |
| 10 | 4 | 26.0 |  | -0.10 |
| 20 | -4 | 26.7 |  | -0.35 |
| 20 | -2 | 31.9 |  | 4.85 |
| 20 | 0 | 34.0 | 27.05 | 6.95 |
| 20 | 2 | 32.1 |  | 5.05 |
| 20 | 4 | 27.4 |  | 0.35 |
| 30 | -4 | 27.7 |  | 0.10 |
| 30 | -2 | 35.7 |  | 8.10 |
| 30 | 0 | 38.6 | 27.60 | 11.00 |
| 30 | 2 | 35.6 |  | 8.00 |
| 30 | 4 | 27.5 |  | -0.10 |
| 40 | -4 | 28.6 |  | -0.20 |
| 40 | -2 | 40.0 |  | 11.20 |
| 40 | 0 | 43.0 | 28.80 | 14.20 |
| 40 | 2 | 39.3 |  | 10.50 |
| 40 | 4 | 29.0 |  | 0.20 |

**Thermal Conductivity**

The thermal conductivity is determined by calculating the required regression sums and combining with the appropriate heat sink dimensions:

$l$ = *Heated Length/Span of Heat Sink* = 8 *in*
$w$ = *Width of Heat Sink* = 5 *in*
$t$ = *Thickness of Heat Sink* = 0.125 *in*

The following data points are obtained from Table 7-3:

$P_i$ = *Total Module Power for Data Point i*
$x_i$ = *Distance from Center*
$T_i$ = *Measured Temperature Rise above Edge at $x_i$ from Center*

Evaluation of the data points produces the following regression sums:

$$\Sigma(P_i^2) = 15,000 \ W^2$$
$$\Sigma(P_i^2 x_i^2) = 120,000 \ W^2\text{-}in^2$$
$$\Sigma(P_i^2 x_i^4) = 1,632,000 \ W^2\text{-}in^4$$
$$\Sigma(T_i \ P_i) = 2,673 \ °C\text{-}W$$
$$\Sigma(T_i \ P_i x_i^2) = 6,400°C\text{-}W\text{-}in^2$$

$$k_{hs} = \frac{l^4 \Sigma(P_i^2) - 8l^2\Sigma(P_i^2 x_i^2) + 16\Sigma(P_i^2 x_i^4)}{8lwt\left(l^2\Sigma(T_i \ P_i) - 4\Sigma(T_i \ P_i x_i^2)\right)}$$

$$k_{hs} = \frac{(8in)^4 15000 \ W^2 - 8(8in)^2 120000 W^2\text{-}in^2 + 16 x 1632000 W^2\text{-}in^4}{8 x 8in x 5in x 0.125in\left((8in)^2 2673° C\text{-}W - 4 x 6400° C\text{-}W\text{-}in^2\right)} \quad (7\text{-}18)$$

$$k_{hs} = 4.487\frac{W}{in\text{-}°C} = 4.487\frac{W}{in\text{-}°C} x 39.37\frac{in}{m} = 177\frac{W}{m\text{-}°C}$$

Thus the thermal conductivity of the heat sink is *177 W/(m-°C)* which is close to that of aluminum (see 3.7.4).

## 7.3.3   Vibration Data Analysis

**Background**

Vibration measurements are typically conducted by mounting the equipment to be tested in a fixture (see Figure 7-7) with accelerometers mounted in various locations. The fixture/hardware assembly with the accelerometers is then mounted on the head of a electrodynamic vibration exciter for testing. The accelerometers measure absolute acceleration but the stresses developed in hardware depend

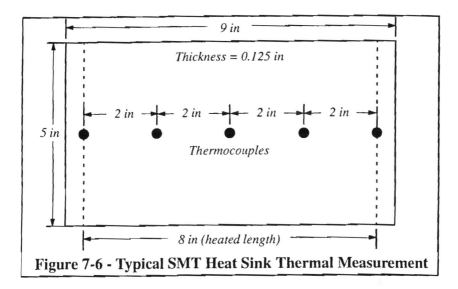

**Figure 7-6 - Typical SMT Heat Sink Thermal Measurement**

upon the relative deflection. Because of the phase relationships between the excitation and response, the relative deflection can not be obtained by subtracting the input vibration level from the response vibration level. Relative vibration response can be obtained by determining the complex response transfer function (complex response over reference) from the vibration analysis system.

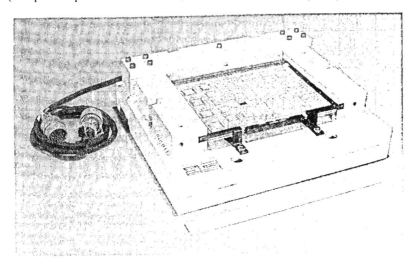

**Figure 7-7 - Typical Module Vibration Configuration**
(Courtesy of Lockheed Martin Control Systems.)

**Relative Deflection for Sinusoidal Vibration**

For sinusoidal vibration, the relative deflection is determined by subtracting unity (real) from the complex response transfer function and performing subsequent calculations as follows:

$$R_{re}(f) = A(f)\cos(\phi(f)) - 1 \tag{7-19}$$

$$R_{im}(f) = A(f)\sin(\phi(f)) \tag{7-20}$$

$$R(f) = \sqrt{(R_{re}(f))^2 + (R_{im}(f))^2} \tag{7-21}$$

$$\theta(f) = \tan^{-1}\left(\frac{R_{im}(f)}{R_{re}(f)}\right) \tag{7-22}$$

$$\delta(f) = \frac{a(f)\ R(f)}{(2\pi f)^2} \tag{7-23}$$

$A(f)$ = Amplitude of Response Transfer Function (response vs. input)
$\phi(f)$ = Phase of Response Transfer Function (response vs. input)
$R_{re}(f)$ = Real Portion of Relative Response Transfer Function
$R_{im}(f)$ = Imaginary Portion of Relative Response Transfer Function
$R(f)$ = Amplitude of Relative Response Transfer Function
$\theta(f)$ = Phase of Relative Response Transfer Function
$\delta(f)$ = Relative Deflection
$a(f)$ = Sinusoidal Input Acceleration
$f$ = Frequency

It should be noted that the preceding equations are frequency-dependent so the evaluation must be conducted at each frequency to properly develop the relative deflection response. It is also important to realize that the response transfer function will be different for various locations on the structure so it is important to properly record the location of the accelerometers used to develop the response curves. Multiple response transfer functions will usually be necessary to develop modeshapes.

**Relative Deflection for Random Vibration**

For random vibration, the relative deflection is determined by integrating the relative response over the frequency range:

$$R_{re}(f) = A(f) \cos(\phi(f)) - 1 \tag{7-19}$$

$$R_{im}(f) = A(f) \sin(\phi(f)) \tag{7-20}$$

$$R(f) = \sqrt{(R_{re}(f))^2 + (R_{im}(f))^2} \tag{7-21}$$

$$g'_{RMS}{}^2 = \int_0^\infty PSD(f) \, (R(f))^2 \, df \tag{4-29}$$

$$g'_{RMS} = \sqrt{\int_0^\infty PSD(f) \left((R_{re}(f))^2 + (R_{im}(f))^2\right) df} \tag{7-24}$$

$$\delta'_{RMS}{}^2 = \int_0^\infty PSD(f) \left(\frac{R(f)}{(2\pi f)^2}\right)^2 df \tag{4-30}$$

$$\delta'_{RMS} = \sqrt{\int_0^\infty PSD(f) \left(\frac{(R_{re}(f))^2 + (R_{im}(f))^2}{(2\pi f)^4}\right) df} \tag{7-25}$$

$A(f) =$ *Amplitude of Response Transfer Function (response vs. input)*
$\phi(f) =$ *Phase of Response Transfer Function (response vs. input)*
$R_{re}(f) =$ *Real Portion of Relative Response Transfer Function*
$R_{im}(f) =$ *Imaginary Portion of Relative Response Transfer Function*
$R(f) =$ *Amplitude of Relative Response Transfer Function*
$g'_{RMS} =$ *Relative RMS Acceleration*
$PSD(f) =$ *Input Power Spectral Density*
$\delta'_{RMS} =$ *RMS Deflection*

As was noted before, care should be taken to ensure that the proper conversion for deflection and acceleration are used when the power spectral density is expressed in terms of $g^2/Hz$.

## Stresses and Modeshapes from Test Data

Modeshapes may be obtained from test data by generating a finite-element model of the equipment under test and applying the deflections given by equation 7-23 to the appropriate locations. Once the finite-element model is solved using the given deflections, the modeshapes may be plotted and/or visualized depending upon the capabilities of the finite-element code. Stresses may also be obtained from the model if the construction and material properties closely match the actual situation. It is important to note that the stress obtained from the model in the immediate area of an applied deflection may not be correct.

### 7.3.4   Typical Vibration Data Analysis

Sinusoidal vibration testing conducted at 2 g (peak) input resulted in the amplitude and phase responses shown in Figure 7-8. Determine the relative deflection at the resonant frequency.

**Relative Deflection**

From observing Figure 7-8, it can be seen that the response to the 2-g input is 20 g at the 100-Hz resonance and the phase angle is -90°. Based on this information, the relative deflection is calculated as follows:

$$f = Frequency = 100\,Hz$$
$$A(f) = Amplitude\ of\ Response\ Transfer\ Function = \frac{20\,g}{2\,g} = 10$$
$$\phi(f) = Phase\ of\ Response\ Transfer\ Function = -90°$$
$$a(f) = Sinusoidal\ Input\ Acceleration = 2\,g = 2x386.4\frac{in}{sec^2} = 772.8\frac{in}{sec^2}$$

$$R_{re}(f) = A(f)\cos(\phi(f)) - 1 = 10\cos(-90°) - 1 = -1 \qquad (7\text{-}19)$$

$$R_{im}(f) = A(f)\sin(\phi(f)) = 10\sin(-90°) = -10 \qquad (7\text{-}20)$$

$$R(f) = \sqrt{(R_{re}(f))^2 + (R_{im}(f))^2} = \sqrt{(-1)^2 + (-10)^2} = 10.05 \qquad (7\text{-}21)$$

$$\delta(f) = \frac{a(f)\ R(f)}{(2\pi f)^2} = \frac{772.8\frac{in}{sec^2}x10.05}{(2\pi x100Hz)^2} = \textbf{0.0197}\ \textbf{\textit{in}} \qquad (7\text{-}23)$$

So the relative deflection at resonance for the system shown in Figure 7-8 is *0.0197 in.*

### 7.3.5   Life Data Analysis

**Background**

One of the typical characteristics of life data is the variation or scatter in the results. Selection of an appropriate sample size (see 7.2.4) can help to characterize

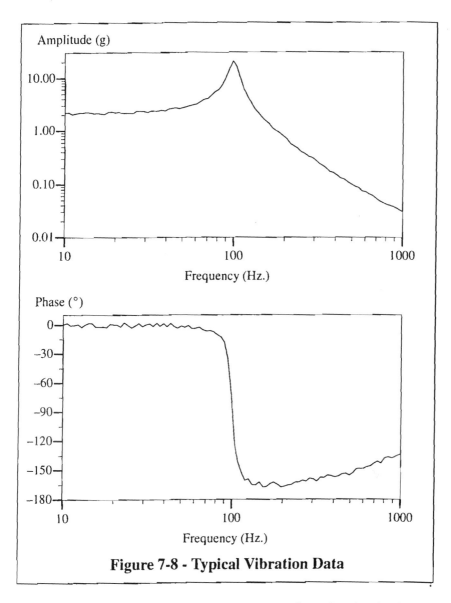

**Figure 7-8 - Typical Vibration Data**

this scatter, but it is important that the data be correctly analyzed to develop the appropriate reliability correction factors. One popular method for analyzing fatigue data is to simply compute the mean cycles to fail as the failures are accumulated. This method is only accurate when all of the devices under test have failed. This is illustrated in Table 7-4 where simulated failure data for nine devices is analyzed [69]. As subsequent items fail, the mean increases from 407 to 877. A method which is well-suited to analysis of life data where not all of the test samples have failed is known as hazard analysis.

| Table 7-4 - Running Mean of Failure Data | |
|:---:|:---:|
| **Cycles to Fail** | **Running Mean** |
| 407 | 407 |
| 549 | 478 |
| 662 | 539 |
| 764 | 595 |
| 864 | 649 |
| 966 | 702 |
| 1077 | 755 |
| 1210 | 812 |
| 1396 | 877 |
| Note: Simulated failure data (Weibull distribution) | |

**Hazard Analysis [68]**

<u>Background</u>

Hazard is defined as the conditional probability of failure given that a failure has not occurred [68]. The hazard function is expressed mathematically as follows:

$$h(x) = \frac{f(x)}{1 - F(x)} \qquad (7\text{-}26)$$

$$f(x) = \frac{d}{dx} F(x) \qquad (7\text{-}27)$$

The cumulative hazard is calculated as follows:

$$H(x) = \int_0^x h(x)dx \qquad (7\text{-}28)$$

$$H(x) = -\ln(1 - F(x)) \tag{7-29}$$

$h(x)$ = Hazard Function
$x$ = Randomly Distributed Variable
$f(x)$ = Probability Density Function
$F(x)$ = Cumulative Failure Probaility
$H(x)$ = Cumulative Hazard

For a Weibull distribution which is used for many fatigue situations (including solder fatigue) the cumulative hazard is calculated as follows:

$$F(x) = 1 - e^{-\left(\frac{x}{a'}\right)^{\beta}} \tag{7-30}$$

$$H(x) = \left(\frac{x}{a'}\right)^{\beta} \tag{7-31}$$

The effective mean (expected value) of the Weibull distribution is given by the following [70]:

$$\bar{x} = a' \, \Gamma\left(1 + \frac{1}{\beta}\right) \tag{7-32}$$

$F(x)$ = Cumulative Failure Probaility
$x$ = Weibull Distributed Variable
$a'$ = Weibull Scale Parameter
$\beta$ = Weibull Shape Parameter
$H(x)$ = Cumulative Hazard
$\bar{x}$ = Expected Value of Weibull Distribution
$\Gamma(t)$ = Gamma Function = $\displaystyle\int_{0}^{\infty} e^{-s} \, s^{t-1} \, ds$

Similar relationships for other probability distribution functions may be found in appropriate literature [68], [70].

Analysis Method

Hazard analysis provides a method for predicting the reliability distribution before all of the items have failed.  Step by step, the hazard analysis is performed as follows [68]:

1.  Sort the failure data from lowest to highest including any unfailed items

2.  Number the list in reverse order to determine the number of items with life greater than or equal to the item

3.  Calculate the percent hazard for the failed items only by dividing 100 by the number from step 2

4.  Sum the hazard values starting at the first failure to determine the cumulative hazard

5.  Plot the failure data vs. cumulative hazard on log-log paper

6.  Obtain the Weibull parameters from the regression line drawn through the data ($\beta$ = 1/Slope, $\alpha'$ = Intercept at 100% cumulative hazard)

Hazard analysis of the simulated failure data described above is summarized in Table 7-5 and Figure 7-9 [69]. Parameter estimates for alpha, beta, and the mean are based upon the simulated failures occurring before the data point (analogous to the running mean). The error between the equivalent mean using running mean and hazard analysis is summarized in Table 7-6. It can be seen from the results that the hazard analysis method converges on estimated parameters rapidly with only a few items failed.

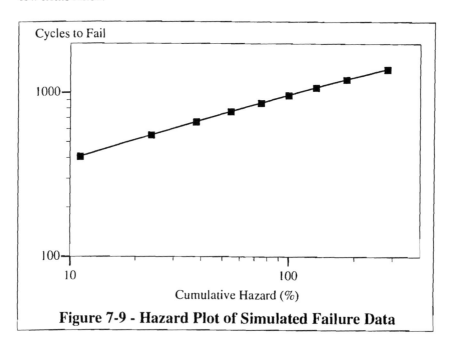

**Figure 7-9 - Hazard Plot of Simulated Failure Data**

| Table 7-5 - Hazard Analysis of Simulated Failure Data | | | | | | |
|---|---|---|---|---|---|---|
| Cycles to Fail | Number* | Hazard | Cumulative Hazard | Weibull Beta | Weibull Alpha | Equivalent Mean |
| 407 | 9 | 11.1 | 11.1 | Insufficient Data | | |
| 549 | 8 | 12.5 | 23.6 | 2.51 | 975 | 865 |
| 662 | 7 | 14.3 | 37.9 | 2.52 | 974 | 864 |
| 764 | 6 | 16.7 | 54.6 | 2.52 | 973 | 863 |
| 864 | 5 | 20.0 | 74.6 | 2.53 | 971 | 862 |
| 966 | 4 | 25.0 | 99.6 | 2.53 | 970 | 861 |
| 1077 | 3 | 33.3 | 132.9 | 2.54 | 967 | 859 |
| 1210 | 2 | 50.0 | 182.9 | 2.56 | 964 | 856 |
| 1396 | 1 | 100.0 | 282.9 | 2.60 | 958 | 851 |

\* Number of items with failure times greater or equal to the failed item (see text)

Note: Parameter estimates for alpha, beta, and equivalent mean are based only upon simulated failures occurring before the data point

## 7.3.6 Typical Life Data Analysis

Thermal cycle life testing of 20 parts for 3,000 cycles resulted in the failure data shown in Table 7-7. Determine the Weibull reliability parameters for the part.

**Hazard Analysis**

Sorting and Numbering of Data

The initial step in hazard analysis is to sort the data in numerical order (including any unfailed items) from lowest to highest, and number the data in reverse order. This is illustrated in Table 7-8.

Calculation of Percent Hazard and Cumulative Hazard

The percent hazard is determined for the failed items by dividing 100 by the failure number from Table 7-8. Once the percent hazard is determined, the cumulative hazard is determined by summing the hazard starting at the first failure as shown in Table 7-9.

| Table 7-6 - Comparison of Simulated Failure Data Analysis | | | |
| Number of Data Points | Running Mean | Equivalent Hazard Mean | Error (%) |
|:---:|:---:|:---:|:---:|
| 1 | 407 | Insufficient Data | |
| 2 | 478 | 865 | -44.7 |
| 3 | 539 | 864 | -37.6 |
| 4 | 595 | 863 | -31.0 |
| 5 | 649 | 862 | -24.7 |
| 6 | 702 | 861 | -18.5 |
| 7 | 755 | 859 | -12.1 |
| 8 | 812 | 856 | -5.1 |
| 9 | 877 | 851 | 3.1 |

Plotting of Cumulative Hazard Data

The cycles to fail is plotted vs. the cumulative hazard as shown in Figure 7-10. Since the plotted data is a straight line on a log-log scale, the data fits a Weibull distribution.

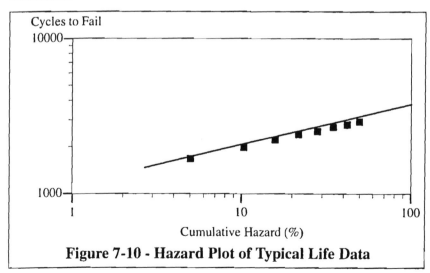

Cycles to Fail

Cumulative Hazard (%)

**Figure 7-10 - Hazard Plot of Typical Life Data**

| Table 7-7 - Typical Life Data | | |
|---|---|---|
| Sample | Cycles | Status |
| 1 | 3,000 | Unfailed |
| 2 | 2,948 | Failed |
| 3 | 2,231 | Failed |
| 4 | 3,000 | Unfailed |
| 5 | 1,685 | Failed |
| 6 | 1,991 | Failed |
| 7 | 2,718 | Failed |
| 8 | 3,000 | Unfailed |
| 9 | 3,000 | Unfailed |
| 10 | 2,547 | Failed |
| 11 | 3,000 | Unfailed |
| 12 | 3,000 | Unfailed |
| 13 | 3,000 | Unfailed |
| 14 | 3,000 | Unfailed |
| 15 | 3,000 | Unfailed |
| 16 | 3,000 | Unfailed |
| 17 | 2,425 | Failed |
| 18 | 3,000 | Unfailed |
| 19 | 2,817 | Failed |
| 20 | 3,000 | Unfailed |

| Table 7-8 - Sorted and Numbered Typical Life Data | | | |
|:---:|:---:|:---:|:---:|
| **Sample** | **Cycles** | **Status** | **Number** |
| 5 | 1685 | Failed | 20 |
| 6 | 1991 | Failed | 19 |
| 3 | 2231 | Failed | 18 |
| 17 | 2425 | Failed | 17 |
| 10 | 2547 | Failed | 16 |
| 7 | 2718 | Failed | 15 |
| 19 | 2817 | Failed | 14 |
| 2 | 2948 | Failed | 13 |
| 1 | 3000 | Unfailed | 12 |
| 4 | 3000 | Unfailed | 11 |
| 8 | 3000 | Unfailed | 10 |
| 9 | 3000 | Unfailed | 9 |
| 11 | 3000 | Unfailed | 8 |
| 12 | 3000 | Unfailed | 7 |
| 13 | 3000 | Unfailed | 6 |
| 14 | 3000 | Unfailed | 5 |
| 15 | 3000 | Unfailed | 4 |
| 16 | 3000 | Unfailed | 3 |
| 18 | 3000 | Unfailed | 2 |
| 20 | 3000 | Unfailed | 1 |

| Table 7-9 - Calculation of Hazard and Cumulative Hazard | | | |
|---|---|---|---|
| Cycles | Number | Hazard | Cumulative Hazard |
| 1685 | 20 | 5.00% | 5.0% |
| 1991 | 19 | 5.26% | 10.3% |
| 2231 | 18 | 5.56% | 15.8% |
| 2425 | 17 | 5.88% | 21.7% |
| 2547 | 16 | 6.25% | 28.0% |
| 2718 | 15 | 6.67% | 34.6% |
| 2817 | 14 | 7.14% | 41.8% |
| 2948 | 13 | 7.69% | 49.5% |

Weibull Parameter Estimates

The Weibull parameters are estimated by conducting a linear regression on the log of cycles to fail vs. the log of the cumulative hazard as follows:

$n = Number\ of\ Data\ Points = 8$

Letting:

$x_i = log_{10}\ of\ Cumulative\ Hazard\ for\ Data\ Point\ i$

$y_i = log_{10}\ of\ Cycles\ to\ Fail\ for\ Data\ Point\ i$

The required regression sums are calculated as follows:

$\Sigma(x_i) = -5.4534$

$\Sigma(y_i) = 27.018$

$\Sigma(x_i^2) = 4.5079$

$\Sigma(x_iy_i) = -18.224$

$$m = \frac{n\Sigma(x_iy_i) - \Sigma(x_i)\Sigma(y_i)}{n\Sigma(x_i^2) - (\Sigma(x_i))^2}$$

$$m = \frac{8(-18.224) - (-5.4534)27.018}{8(4.5079) - (-5.4534)^2} = 0.245$$

(7-33)

$$b = \frac{\Sigma(y_i) - m\Sigma(x_i)}{n} = \frac{27.018 - 0.245(-5.4534)}{8} = 3.544 \qquad (7\text{-}34)$$

$$\beta = \frac{1}{m} = \frac{1}{0.245} = 4.08 \qquad (7\text{-}35)$$

$$a' = 10^b = 10^{3.544} = 3,499 \ cycles \qquad (7\text{-}36)$$

So the Weibull alpha for the above data is *3,499 cycles* and the shape parameter (beta) is *4.08*. These Weibull parameters may be used to estimate the expected value of the distribution (effective mean) as follows [70]:

$a' = $ *Weibull Scale Parameter* $ = 3,499 \ cycles$
$\beta = $ *Weibull Shape Parameter* $ = 4.08$

$$\Gamma\left(1 + \frac{1}{\beta}\right) = \Gamma\left(1 + \frac{1}{4.08}\right) = 0.9074$$

$$\bar{x} = a' \ \Gamma\left(1 + \frac{1}{\beta}\right) = 3,499 \ cycles \ \Gamma\left(1 + \frac{1}{4.08}\right) = 3,175 \ cycles \qquad (7\text{-}32)$$

### 7.3.7    Thermal Expansion Data Analysis

**Background**

As was mentioned earlier in this chapter, the difference in temperature-induced strain between a test sample and a reference sample (see Figure 7-11) may be used to determine the coefficient of thermal expansion (CTE) rates [64],[65]. Typically, titanium silicate is used as a reference material since it has a thermal expansion rate close to zero over normal electronic equipment operating temperatures. However, any material with a well-known expansion rate over temperature may be used. Since temperature has an effect on resistance for both the strain gage and lead wires, care must be taken to ensure that the length of the lead wires inside the chamber is nearly the same for the test sample and the reference sample. Once the equipment is assembled, the chamber can be taken over the required temperature range and the strain measured at each temperature. It must be noted that sufficient time must be given for the test sample and reference sample to reach the same temperature. It is also important that the strain gages have the same temperature-induced apparent strain characteristics. This typically means that the gages must be from the same manufactured lot.

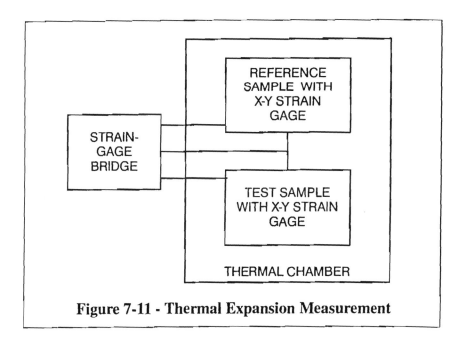

**Figure 7-11 - Thermal Expansion Measurement**

## Analysis

Analysis of the resulting data is relatively simple if a single expansion rate is desired over a relatively large temperature range. In this case the CTE is the difference in strain over the difference in temperature:

$$\alpha_t - \alpha_r = \frac{\epsilon_h - \epsilon_l}{T_h - T_l} \qquad (7\text{-}37)$$

$\alpha_{t,r}$ = CTE of Test, Reference Samples
$\epsilon_{h,l}$ = Measured Relative Strain at High, Low Temperatures
$T_{h,l}$ = High, Low Test Temperature

Drift in the expansion measurement can produce a phenomenon similar to that encountered in transient thermal measurements (see Figure 7-2). This situation is especially problematic if the change in CTE over temperature is to be measured. In this situation, good results may be obtained by fitting the expansion data to a polynomial and calculating the derivative to determine the CTE. Use of a cubic equation for expansion allows for a quadratic variation in CTE, and is calculated as follows:

$$\epsilon = a_0 + a_1T + a_2T^2 + a_3T^3 \tag{7-38}$$

$$a_t - a_r = a_1 + 2a_2T + 3a_3T^2 \tag{7-39}$$

Coefficients of the polynomial are determined by simultaneously solving the following equations:

$$na_0 + \Sigma(T_i)a_1 + \Sigma(T_i^2)a_2 + \Sigma(T_i^3)a_3 = \Sigma(\epsilon_i) \tag{7-40}$$

$$\Sigma(T_i)a_0 + \Sigma(T_i^2)a_1 + \Sigma(T_i^3)a_2 + \Sigma(T_i^4)a_3 = \Sigma(\epsilon_iT_i) \tag{7-41}$$

$$\Sigma(T_i^2)a_0 + \Sigma(T_i^3)a_1 + \Sigma(T_i^4)a_2 + \Sigma(T_i^5)a_3 = \Sigma(\epsilon_iT_i^2) \tag{7-42}$$

$$\Sigma(T_i^3)a_0 + \Sigma(T_i^4)a_1 + \Sigma(T_i^5)a_2 + \Sigma(T_i^6)a_3 = \Sigma(\epsilon_iT_i^3) \tag{7-43}$$

$\epsilon$ = *Relative Expansion (strain)*

$T$ = *Temperature*

$a_{0,1,2,3}$ = *Coefficients of Polynomial*

$a_{t,r}$ = *CTE of Test, Reference Samples*

$n$ = *Number of Data Points*

$T_i$ = *Temperature at Data Point i*

$\epsilon_i$ = *Measured Relative Expansion at Data Point i*

It is important that enough data points are available so that the temperature variation in expansion can be properly defined. It is also important that the polynomial is of sufficient order to properly represent this temperature variation. Plotting the actual expansion data and the expansion predicted by the polynomial on the same axes may help to determine if the actual data and that predicted by the polynomial are consistent.

### 7.3.8    Typical Thermal Expansion Data Analysis

Thermal expansion testing conducted on a material results in the expansion vs. temperature data shown in Table 7-10 (measured against a material with a CTE of 0 ppm/°C). Determine the thermal expansion as a function of temperature for the material.

**Regression of Expansion Data**

Using a cubic representation for the expansion data, the following regression representation is determined:

$$\epsilon = a_0 + a_1T + a_2T^2 + a_3T^3 \tag{7-38}$$

$n$ = *Number of Data Points* = 10

The required regression sums are calculated as follows:

$$\Sigma(T_i) = 350°C$$
$$\Sigma(T_i^2) = 45,250°C^2$$
$$\Sigma(T_i^3) = 3,893,750°C^3$$
$$\Sigma(T_i^4) = 4.50936x10^8°C^4$$
$$\Sigma(T_i^5) = 4.85155x10^{10}°C^5$$
$$\Sigma(T_i^6) = 5.64544x10^{12}°C^6$$
$$\Sigma(\epsilon_i) = 2,040\ ppm$$
$$\Sigma(\epsilon_i T_i) = 658,280\ ppm°C$$
$$\Sigma(\epsilon_i T_i^2) = 51,118,200\ ppm°C^2$$
$$\Sigma(\epsilon_i T_i^3) = 6.48782x10^9\ ppm°C^3$$

| Table 7-10 - Typical Thermal Expansion Data | |
|:---:|:---:|
| Temperature (°C) | Expansion (ppm) |
| -55 | -1359 |
| -35 | -1032 |
| -15 | -671 |
| 5 | -367 |
| 25 | 39 |
| 45 | 313 |
| 65 | 735 |
| 85 | 1054 |
| 105 | 1462 |
| 125 | 1866 |

Using the regression sums in the simultaneous equations (units omitted for clarity):

$$na_0 + \Sigma(T_i)a_1 + \Sigma(T_i^2)a_2 + \Sigma(T_i^3)a_3 = \Sigma(\epsilon_i)$$

$$10a_0 + 350a_1 + 45,250a_2 + 3,893,750a_3 = 2,040 \tag{7-40}$$

$$\Sigma(T_i)a_0 + \Sigma(T_i^2)a_1 + \Sigma(T_i^3)a_2 + \Sigma(T_i^4)a_3 = \Sigma(\epsilon_i T_i) \tag{7-41}$$

$$350a_0 + 45,250a_1 + 3,893,750a_2 + 4.50936x10^8 a_3 = 658,280$$

$$\Sigma(T_i^2)a_0 + \Sigma(T_i^3)a_1 + \Sigma(T_i^4)a_2 + \Sigma(T_i^5)a_3 = \Sigma(\epsilon_i T_i^2) \tag{7-42}$$

$$45,250a_0 + 3,893,750a_1 + 4.50936x10^8 a_2 + 4.85155x10^{10} a_3 = 51,118,200$$

$$\Sigma(T_i^3)a_0 + \Sigma(T_i^4)a_1 + \Sigma(T_i^5)a_2 + \Sigma(T_i^6)a_3 = \Sigma(\epsilon_i T_i^3)$$

$$3,893,750a_0 + 4.50936x10^8 a_1 + 4.85155x10^{10} a_2 + \tag{7-43}$$
$$5.64544x10^{12} a_3 = 6.48782x10^9$$

The polynomial coefficients are determined as follows:

$$a_0 = \text{-}427.559 \; ppm \qquad\qquad a_2 = 2.625x10^{-3} \frac{ppm}{°C^2}$$

$$a_1 = 16.973 \frac{ppm}{°C} \qquad\qquad a_3 = 6.581x10^{-5} \frac{ppm}{°C^3}$$

A comparison of the data with the calculated regression is shown in Figure 7-12.

## CTE Determination

The coefficient of thermal expansion is determined by taking the derivative of equation 7-38 as follows:

$$a_r = CTE \; of \; Reference \; Sample = 0 \; \frac{ppm}{°C}$$

Polynomial coefficients:

$$a_0 = \text{-}427.559 \; ppm \qquad\qquad a_2 = 2.625x10^{-3} \frac{ppm}{°C^2}$$

$$a_1 = 16.973 \frac{ppm}{°C} \qquad\qquad .\, a_3 = 6.581x10^{-5} \frac{ppm}{°C^3}$$

$$a_t - a_r = a_1 + 2a_2 T + 3a_3 T^2 \tag{7-39}$$

$$a_t - 0 \frac{ppm}{°C} = 16.973 \frac{ppm}{°C} + 2x2.625x10^{-3} \frac{ppm}{°C^2} T + 3x6.581x10^{-5} \frac{ppm}{°C^3} T^2$$

$$a_t = 16.973 \frac{ppm}{°C} + 5.25x10^{-3} \frac{ppm}{°C^2} T + 1.974x10^{-4} \frac{ppm}{°C^3} T^2$$

The resulting CTE vs. temperature is shown in Table 7-11 and Figure 7-13. The data indicated by the squares on Figure 7-13 represent the CTE obtained by the difference in expansion over the temperature change for consecutive data points, illustrating the ability of the regression to smooth out random variations due to drift, etc. The slight increase in CTE at low temperatures may be a characteristic of the regression approach in conjunction with random data variations. This may possibly be eliminated by increasing the polynomial order or increasing the number of data points. This helps to emphasize that the polynomial should not be used to extrapolate beyond the measured temperature range.

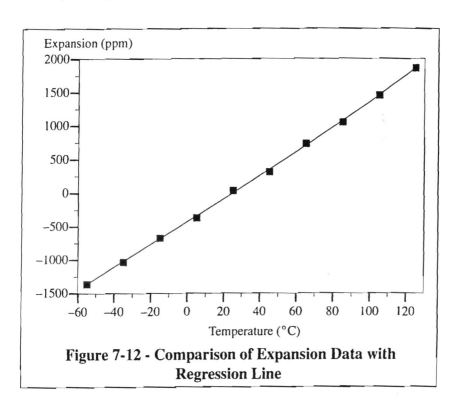

**Figure 7-12 - Comparison of Expansion Data with Regression Line**

# 7.4   VERIFICATION

Any results from test data analysis should be verified to help avoid any inaccuracies that might arise. This verification may be by alternate testing methods, analysis, or other approaches as appropriate. For information on verification, see Chapter 8.

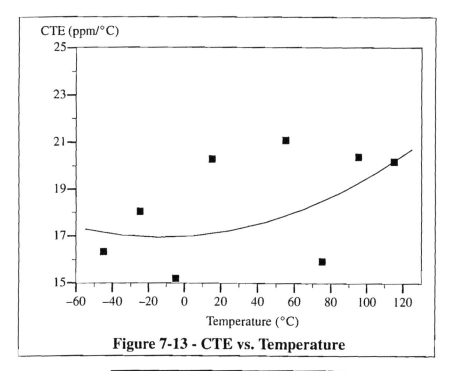

**Figure 7-13 - CTE vs. Temperature**

| Table 7-11 - Calculated CTE vs. Temperature | |
| --- | --- |
| Temperature (°C) | CTE (ppm/°C) |
| -55 | 17.3 |
| -35 | 17.0 |
| -15 | 16.9 |
| 5 | 17.0 |
| 25 | 17.2 |
| 45 | 17.6 |
| 65 | 18.1 |
| 85 | 18.8 |
| 105 | 19.7 |
| 125 | 20.7 |

# 7.5    TEST DATA ANALYSIS CHECKLIST

## 7.5.1    Applicable to Test Planning

☐ Does the test approach adequately represent the actual situation?

☐ Does the instrumentation system have sufficient resolution, sensitivity, and repeatability to obtain the desired results?

☐ Is the accuracy of the instrumentation system consistent with the desired accuracy?

☐ Do the test levels avoid significant non-linearities?

☐ Are the test levels within the capabilities of the component materials?

☐ Is the relationship between test level and expected use for life testing such that deviation from the fatigue curve is insignificant or is it accountable?

☐ Is the sample size large enough to account for variations in test data?

## 7.5.2    Applicable to Thermal Data Analysis

☐ Has a regression been used where there is significant variation in test data?

☐ Are the thermal sensors in a SMT heat sink analysis within the heated length of the heat sink?

☐ Are the edge temperatures in the SMT heat sink equal?

☐ Do the thermal boundary conditions (power, temperature) of the test case match the actual situation?

☐ Where applicable, does the surrounding fluid velocity and radiation environment of the test case match the actual situation?

☐ For transient thermal measurements, does the thermal mass of the test case match the actual configuration?

## 7.5.3    Applicable to Vibration Data Analysis

☐ Has proper consideration been given to phase relationships in the determination of relative deflection?

☐ Has relative deflection been used to visualize modeshapes?

## 7.5.4    Applicable to Life Analysis

☐ Has life data been analyzed with proper consideration given to unfailed items?

### 7.5.5    Applicable to Expansion Measurement

☐   Is the expansion rate of the reference material known?

☐   Do the strain gages on the test and reference samples have identical apparent strain characteristics?

☐   Is the length of lead wire within the thermal chamber the same?

☐   Is sufficient time given during the thermal testing for the test and reference samples to reach the same temperature?

☐   Are enough data points available to properly characterize the temperature variation in expansion?

☐   Is the order of the polynomial consistent with the variation in expansion over temperature?

☐   Is the resulting expansion calculation within the measured temperature range?

### 7.5.6    Other Data Required

In addition to the data measured as part of a test, the following data is typically required for various data analyses. Typical units are provided for convenience, and care must be taken to ensure that the system of units is consistent with the material properties and model dimensions.

**PWB/Module Thermal Data Analysis**

☐   Length (heated), width, thickness of SMT heat sink (cm, m, in, etc.)

☐   Locations of thermal sensors along length of SMT heat sink (cm, in, etc.)

**Thermal Expansion Data Analysis**

☐   CTE of reference sample over temperature (ppm/°C, in/in/°F, etc.)

## 7.6    REFERENCES

[64]   Finke & Heberling; "Determination of Thermal-Expansion Characteristics of Metals Using Strain Gages", Experimental Mechanics, April 1978, pp 155-158

[65]   Measurements Group; "How to Measure Expansion Coefficients with Strain Gages"; R & E Technical Education Newsletter, No. 32, February 1981; Measurements Group, Raleigh, NC

[66] Sonntag & Van Wylen; Introduction to Thermodynamics: Classical and Statistical; John Wiley & Sons, 1971

[67] Book, Stephen; Essentials of Statistics; McGraw-Hill, 1978

[68] Nelson, W.; "Hazard Plotting for Incomplete Failure Data"; Journal of Quality Technology, Volume 1, Number 1, January 1969; pages 27-52

[69] McKeown, S.; "Solder Life Prediction for Leadless Chip Carriers"; Proceedings of the 11th Digital Avionics Systems Conference, October 5-8, 1992; IEEE, Piscataway, NJ

[70] Mann, Schafer, Singpurwalla; Methods for Statistical Analysis of Reliability and Life Data; John Wiley & Sons, 1974

# Chapter 8
# Reporting and Verification

## 8.1  INTRODUCTION

A key to any analysis activity is the communication of results. Typically, the results are documented in a report that states the objective, results, and conclusions of the analysis. Also included are the relevant approach, assumptions, and data references. The verification of any analysis should be an on-going process, but since verification is applicable to a number of analytical techniques, the approaches for verification are included in this chapter.

## 8.2  REPORTING

In many instances, technical reporting is controlled by procedures and/or standards specific to an individual organization. This chapter is not intended to replace or supersede those guidelines, but it is hoped that additional value may be added by the guidelines provided herein. A typical analysis report includes the following:

- A summary section
- The objective of the analysis
- A description of the equipment being analyzed and the conditions of the analysis
- The technical approach used in the analysis
- Any assumptions used
- Description of any testing used in the analysis
- The results of the analysis
- The conclusions and recommendations obtained from the results
- Figures and tables
- References used for the technical approach or data sources
- Appendix (if required)

The material between the summary and the appendix is often referred to as the body of the report.

327

## 8.2.1    <u>Summary</u>

The summary section provides a means for the reader to quickly determine the results of the analysis and the relevance to his or her situation without reading the entire report. Although this is not intended to replace the full report, it can help to provide a concise "bottom line" for those interested in the results that do not require all of the details. Typically a summary includes the following:

- An abstract
- The objective of the analysis
- A summary of the results
- The conclusions and recommendations obtained from the results

It is important that the summary material be self-sufficient so that the full report is not required to understand the summary material. This self-sufficiency is particularly important in organizations where paper is saved by only distributing the summary pages to some individuals. It is also important to note that the body of the report should also be self-sufficient since some readers may skip the summary material and start with the body. [71]

### Abstract

An abstract is particularly useful where a computer database of technical reports is available. In these instances, the abstract can provide a very concise summary of analysis approach, results, and conclusions in a single paragraph. Usually an informative abstract that includes some of the key results and conclusions is preferred over the descriptive abstract which includes only the topics discussed. The main goal of the informative abstract is to incorporate as much useful material as possible without making the abstract excessively long. [71]

Although there is some overlap in the material contained in the abstract and that contained in the summary sections, the use of bulletized lists and sections in the summary material typically makes the summary easier to follow. The use of bulletized lists may produce problems in portability between different computer platforms, but the text-based abstract in single-paragraph form may be displayed on almost any computer system without problems due to differences in the specific computer system.

### Objective

The objective should answer the question, "Why are we doing this analysis?". Stating the objective can help to keep the analysis (and the report) focused on the needs, and help avoid activities that are not relevant. A bulletized list may be helpful in making the objectives easy to follow, especially when the conclusions are to be related to the objectives.

**Results Summary**

The results summary need not provide all of the details of the results that have been obtained from the analysis, but should provide the key points of the results that have been used to obtain the conclusions. The use of figures supporting the key points can be very helpful in clarifying the results.

**Conclusions/Recommendations**

The conclusions provide interpretation of the results and their relationship to the objectives. In some cases, a one-to-one correspondence between objectives and conclusions may be helpful in establishing that the objectives have been met. Recommendations provide suggestions for solutions and any further action that may be required. Conclusions and recommendations should generally be constructive, avoiding negative statements without hiding the facts. [72]

## 8.2.2 Objective

As stated earlier, the objective should answer the question "Why are we doing this analysis?". In most cases, the objective in the body of the report may be an exact copy of the objective used in the summary. This provides self-sufficiency for those who may choose to skip the summary (see 8.2.1). In cases where the events leading up to the analysis are not clear, additional background information may be included to support the objective.

## 8.2.3 Description

This section describes the equipment being analyzed, and provides the material properties and environments that a reader may need to re-construct the analysis at a later date. References should be provided for all of the descriptive information provided in this section including the following (see 8.2.10):

- Environment specifications
- Material specifications
- Vendor data
- Engineering drawings
- Computer data files

Figures and tables may be particularly helpful in conveying the descriptive information contained in this section.

### 8.2.4    Approach

The approach provides the technical basis for the analysis method used. Included in this would be the basic physical principle or field of study relevant to the analysis. In most cases, the appropriate references for the technical approach should be stated (see 8.2.10) although this may not be necessary if generally accepted methods are used or if a detailed description is included in the report. Any supporting material such as equation derivations or software listings may be included in the appendix if too bulky for the main report.

### 8.2.5    Assumptions

This includes any assumptions used in establishing material properties, boundary conditions, environments, or the technical approach. Included in this are any underlying assumptions used in the reference material that can be expected to influence the results. Also included should be any technical approach references that, although consistent with the analysis, have not been verified as part of the analysis or the verification data is not available (see underlying assumptions of 5.2.3 and 6.4.3). Any information that supports the assumptions should be described or referenced to aid the reader in forming his or her own conclusions as to their validity.

### 8.2.6    Testing

This section provides a description of any testing conducted including the test conditions and equipment used. As was the case with the description, sufficient detail should be included to allow the testing to be repeated at a later date if necessary. Figures and tables may help in describing the test configurations and conditions. Any references describing test methods, other sources of data, etc. should be stated.

### 8.2.7    Results

This includes the results of the analysis and/or any testing conducted. Use of figures and tables may help to clarify the presentation of results and help in establishing comparisons. Result material too bulky for inclusion in the main report may be included in the appendix, although any results used in establishing conclusions and recommendations should be included (or at least summarized) in the main report.

### 8.2.8    Conclusions/Recommendations

The conclusions and recommendations provide interpretation of results and suggestions for solutions and further action. In most cases, the conclusions and recommendations in the body of the report may be an exact copy of those used in the summary. As was the case with the objective, repeating the conclusions and recommendations provides self-sufficiency for those who may choose to skip the summary (see 8.2.1).

## 8.2.9    Figures and Tables

Figures and tables may be interspersed in the text of the report, or in separate sections following the main text. Where possible, the figure or table should be placed after its reference in the text. Typically, the preparation of the report is easier if the figures and tables are included in separate sections since this helps to avoid unusual page sizes, but this is largely a matter of personal preference or organizational policy. Large tables of information provided as a matter of record should be included in the appendix. [71]

Typical figures appropriate to analysis reports may include the following:

- Drawings of photographs showing the equipment being analyzed
- Drawings of photographs showing test configurations
- Graphs showing environmental temperature vs. time, vibration vs. frequency, acceleration vs. time, etc.
- Finite-element mesh plots
- Diagrams showing boundary conditions
- Graphs showing variation in results vs. time, temperature, frequency, etc.
- Deflection or modeshape plots
- Contour plots of stress, temperature, etc.

Tables typically included in analysis reports may include:

- Environmental conditions
- Material properties
- Test equipment lists
- Results of analysis and/or testing

Although the use of color may significantly help to clarify the information presented, it must be remembered that monochrome copies are often made of reports and useful information may be lost.

## 8.2.10    References

The reference section is important since it provides support for the credibility of the technical approach and/or testing methods, and traceability for data sources. References are also a professional courtesy in acknowledging the contribution of others in the analytical work. Any reference material that is difficult to trace should be included in the appendix .

**Technical**

Technical references include the theoretical basis for the approach used. This is particularly important in situations, such as solder-life analysis, where different approaches for analyzing a particular problem are used., or where a new approach that is not generally accepted is implemented. Providing reference materials allows the reader to better understand the analytical approach and the underlying assumptions.

**Data Sources**

The accuracy of any analysis is only as good as the input data provided. Including references for data sources allows the reader to verify that the analyzed configuration, material properties, and environmental conditions are current and accurate. Data source references also simplify the reconstruction of an analysis if any changes occur later. All data source references should include the appropriate dates and/or revision levels to provide traceability.

<u>Configuration Drawings</u>

Configuration drawing references provide traceability for the configuration analyzed. Since an analysis typically involves a "snapshot" of a particular configuration, the dates and/or revision level of the documents is particularly important.

<u>Material Properties</u>

Material property references can include the following:

- Test data
- Technical references
- Handbooks
- Vendor data
- Material specifications

It is important to note the implied differences between the various data sources for material properties. Material specifications often provide limits for determining the acceptability of material which is helpful in worst-case analyses. This could be in contrast to vendor data which may represent typical or optimistic values. In some cases, multiple data sources may be helpful in establishing material property limits.

<u>Environmental Conditions</u>

Environmental condition references are typically provided in the customer requirement documents and/or specifications.

<u>Other Data</u>

Other data references that are important to an analysis may include:

- Location of computer data files
- Component data

Component data may include physical data such as overall dimensions and configuration, performance data such as power dissipation or junction-to-case resistance, etc.

## 8.2.11   Appendix

Information included in the appendix includes information too bulky for inclusion in the main report, or that is significant but secondary to the principal message. This information can include:

- Equation derivations
- Test data
- Result information
- Input files and/or source codes
- Reference material that is difficult to trace

It should be noted that the appendix should not be a "dumping ground" for useless information. All appendix material should be specifically related to and referenced by the body of the report. Referencing of appendix material may be easier if the appendix is broken into sections, particularly if a large of amount of information is included. [71]

# 8.3   VERIFICATION

In any analysis it is very important to verify the results throughout the analytical process. Inaccuracies may be introduced into an analysis in any of the following areas:

- Improper analytical approach
- Incorrect source data
- Errors in calculation of results

One of the best methods for providing independent verification is a detailed review of the analysis by peers qualified in the appropriate fields of study. The material provided in this section can facilitate such a peer review.

## 8.3.1    Verification of Approach

### Verification of Theoretical Basis

In any analysis it is important to fully understand the theoretical developments and physical principles used in developing the approach. An analysis should never be the plugging of numbers into equations and calculating the results ("plug and chug"). In many cases there are underlying assumptions or conditions that may or may not be applicable to the particular problem. A proper theoretical understanding is required for any of the analytical methods presented in this work or any others, regardless of the source. In some cases the theoretical basis of the method is understood but the verification data is not available which makes the theoretical basis an assumption (see 8.2.5).

### Verification of Assumptions

Each of the assumptions used in the analysis, including underlying assumptions of the approach, should be examined critically to determine if they are consistent with the specific situation. Any information that supports the assumptions should be included in the report to aid readers in conducting their evaluations.

## 8.3.2    Verification of Source Material

All source material (configuration drawings, material properties, environmental conditions, etc.) should be verified against the reference documents to establish that the correct information was included in the analysis. In some cases (such as component thermal performance) it is important to understand the underlying conditions, environment, or test method used to determine the data (see 3.4.3).

## 8.3.3    Result Verification

One of the keys to successful engineering analysis is to determine if the results obtained make sense. Typically this result verification involves testing and analysis by different methods as summarized in Tables 8-1 and 8-2.

### Comparison with Testing

One of the best methods for verifying an analysis is to conduct testing to verify the results of the analysis. This verification approach may include the following:

- Testing to a specific, as analyzed, requirement
- Modification of the analysis to reflect a specific test case
- Use of similarity analysis to compare differing test and analysis conditions

These approaches for verification by testing are summarized in Table 8-1.

| Table 8-1 - Result Verification by Testing | | |
|---|---|---|
| **Method** | **Advantages** | **Disadvantages** |
| Testing to specific (as analyzed) requirement | • Testing and analysis match requirements<br><br>• Existing analysis may be used | • Testing may not reflect scatter in results<br><br>• Testing may not be practical |
| Analysis of specific test case | • Testing can be designed to reflect available resources | • Testing may not reflect variations in approach<br><br>• Testing may not reflect assumptions<br><br>• New analysis required |
| Similarity comparison of test and analysis | • Testing can be designed to reflect available resources<br><br>• Existing analysis may be used | • Testing may not reflect variations in approach<br><br>• Testing may not reflect assumptions<br><br>• Similarity approach may not properly reflect the comparison |

### Testing to Specific Requirement

In this situation, a test is developed that closely matches the conditions of the analysis. Since the testing matches the analysis, the ability of the hardware to meet the desired objectives and/or requirements is directly verified. If fatigue analysis, tolerance and material property variations, or other situations with scatter in results are involved, it is important that the testing properly assesses the expected scatter in the results.

### Analysis of Specific Test Case

In this method, the analysis is repeated with conditions that match a case that is easily tested. This is especially appropriate in situations where testing of the specific requirement is not possible and/or economical. When this method is used to verify results, it is important that the test case properly reflect the approach and assumptions used in the original analysis.

### Similarity Comparison of Test and Analysis Conditions

A similarity comparison (see 6.6) allows an existing analysis to be compared to a specific test case. This is especially helpful when a new analysis under different conditions is not practical and/or economical. With this method, it is important that the testing properly reflect the approach and assumptions used in the original analysis, and that the similarity comparisons are valid.

**Analysis by Different Methods**

Another approach for verifying an analysis is to conduct another analysis using a different method. This verification method may include varying the finite-element mesh or using a simplified analysis to verify the results (see Table 8-2).

| **Table 8-2 - Result Verification by Analysis** | | |
|---|---|---|
| **Method** | **Advantages** | **Disadvantages** |
| Mesh density variation | • Uses consistent analytical approach | • Only valid with finite-element model |
| | | • Generation of new mesh and analysis may be time-consuming |
| | | • Does not detect errors in finite-element method |
| Comparison with simplified analysis | • Provides independent verification of results (different method) | • Simplified approach may not be possible |
| | | • Simplified approach may be time-consuming |
| | | • Simplifications in analysis may imply errors that do not exist |

Mesh Density Variation

Typically, the accuracy of a finite-element model increases with increasing mesh density. Varying the density of a finite-element mesh allows inaccuracies due to inappropriate mesh size to be detected. In some cases, when generation of a refined mesh is not practical, use of higher-order elements (with more DOFs per node or more nodes per element) may allow the accuracy to be verified without a completely new mesh. Since mesh density variation does not detect errors in the material properties, boundary conditions, or application of the finite-element method, it is important that the finite-element input data be checked carefully.

Comparison with Simplified Analysis

Another method for verifying the results of an analysis is to conduct a simplified analysis (often using classical methods) to verify the results. Since this approach is independent of the original analysis method, errors in original analysis may be detected. It must be noted that variations in material properties and other items common to both the original analysis and the verification will not be detected. To avoid this problem, any input data common to both the original analysis and the verification must be checked carefully. A simplified analysis may be helpful even if verification has already been conducted using a different method.

**Other Considerations**

In addition to verification by testing and analysis by different methods, result verification should also include checking the accuracy of the analysis and the effect of variations in input data. Typically this includes checking for obvious errors and determining the effect of tolerances and material property variations.

Checking for Errors

Even if the approach used in the analysis is valid, errors may be introduced by the following:

- Incorrect copying of equations
- Mathematical errors
- Incorrect copying of source material

To avoid these errors, all equations should be checked against the source to verify that they are copied correctly. This verification is relatively simple where symbolic mathematics software is used but may be difficult for a complex equation in a spreadsheet where the representation of equations is often cryptic. When hand calculations are involved, the results should be checked for mathematical errors. One way of verifying the results is to change the input data to match a case with known results, although it is important to verify that the test case is correct. It also should be reiterated that any source material used in the analysis should be verified (see 8.3.2).

Tolerances and Material Property Variations

Even if the accuracy of an analysis is verified, it is possible that variations due to tolerances or material properties may significantly influence results. This is especially true when the tolerance is a large fraction of the dimension (such as component lead dimensions), or for polymer and composite materials in which the properties are strongly influenced by processing. In cases where tolerances or property variations are significant or the analysis shows a small design margin, the analysis should be repeated for the best-case and worst-case combination of dimensions and properties.

**Typical Analysis Verification**

<u>Original Analysis</u>

An aluminum cantilever beam 1-in square by 10-in long (see Figure 8-1) is subjected to a 20-pound load. A finite-element model of the beam (see Figure 8-2) produces a deflection under the load of 0.0080 in (see Figure 8-3) and maximum bending stress of 1,225 psi (see Figure 8-4). This analysis is to be verified by both mesh variation and simplified analysis.

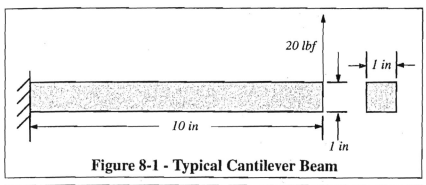

**Figure 8-1 - Typical Cantilever Beam**

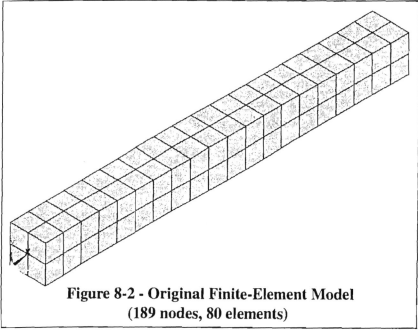

**Figure 8-2 - Original Finite-Element Model**
**(189 nodes, 80 elements)**

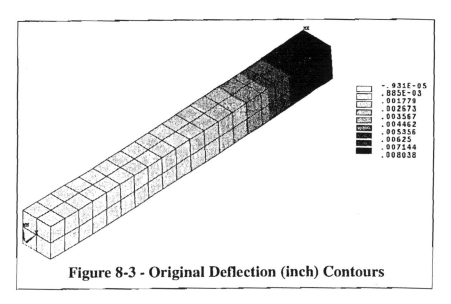

**Figure 8-3 - Original Deflection (inch) Contours**

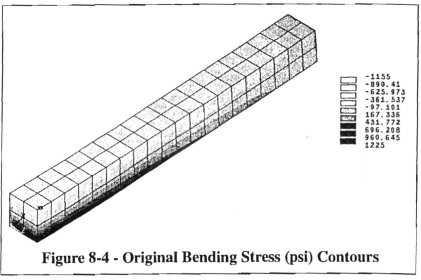

**Figure 8-4 - Original Bending Stress (psi) Contours**

<u>Mesh Variation</u>

The original finite-element model contained 189 nodes and 80 elements (see Figure 8-2). Analysis using an increased mesh density of 1,025 nodes and 640 elements (see Figure 8-5) increased the maximum bending stress by up to 3.8% and increased deflection by 0.1%. A decreased mesh density of 44 nodes and 10 elements reduced the maximum bending stress by up to 7.4% and reduced deflection by 0.1%. So both the increased and decreased mesh densities verify the validity of the original analysis. These results (with specific stress values) are summarized in Table 8-3.

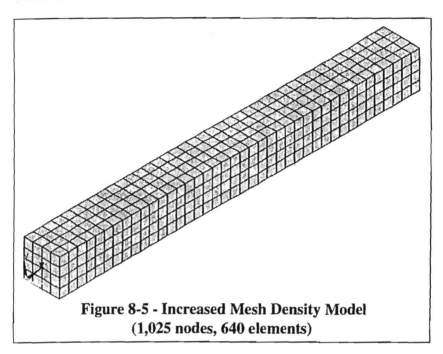

**Figure 8-5 - Increased Mesh Density Model**
**(1,025 nodes, 640 elements)**

Simplified Model

The cantilever beam shown in Figure 8-1 can be modeled using classical beam theory, as follows [73]:

$$\delta = \frac{PL^3}{3EI} \tag{8-1}$$

$$I = \frac{bh^3}{12} \tag{8-2}$$

$$\sigma = \frac{Mc}{I} \tag{8-3}$$

$$c = \frac{h}{2} \tag{8-4}$$

$$M = PL \tag{8-5}$$

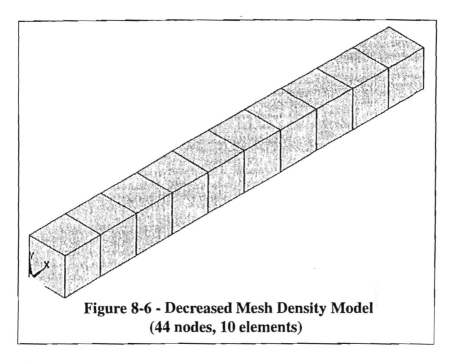

**Figure 8-6 - Decreased Mesh Density Model
(44 nodes, 10 elements)**

$\delta$ = *Maximum Deflection*

$P$ = *Applied Load*

$L$ = *Length of Beam*

$E$ = *Modulus of Elasticity*

$I$ = *Moment of Inertia*

$b$ = *Beam Width*

$h$ = *Beam Height*

$\sigma$ = *Maximum Stress*

$M$ = *Maximum Bending Moment*

$c$ = *Centroid to Extreme Fiber Distance*

Using the specific values appropriate to the configuration, the deflection and stress are calculated:

$b$ = *Beam Width* = 1 *in*

$h$ = *Beam Height* = 1 *in*

$L$ = *Length of Beam* = 10 *in*

$P$ = *Applied Load* = 20 *lbf*

For 6061 aluminum [74]:

$$E = Modulus\ of\ Elasticity = 10x10^6\ psi$$

$$I = \frac{bh^3}{12} = \frac{1in\ x\ (1in)^3}{12} = 0.08333\ in^4 \qquad (8\text{-}2)$$

$$\delta = \frac{PL^3}{3EI} = \frac{20lbf\ x\ (10in)^3}{3x10x10^6psix0.08333in^4} = \mathbf{0.008\ in} \qquad (8\text{-}1)$$

$$c = \frac{h}{2} = \frac{1in}{2} = 0.5\ in \qquad (8\text{-}4)$$

$$M = PL = 20lbfx10in = 200\ in\text{-}lbf \qquad (8\text{-}5)$$

$$\sigma = \frac{Mc}{I} = \frac{200in\text{-}lbf\ x\ 0.5in}{0.08333in^4} = \mathbf{1,200\ psi} \qquad (8\text{-}3)$$

The stress of 1,200 psi is tensile in the bottom of the beam and compressive in the top which is 3.9% higher than that determined by the original analysis. The deflection using simplified analysis is *0.008 in* which is 0.5% lower than shown by the original analysis. These results are summarized in Table 8-3.

| Table 8-3 - Typical Analysis Verification Results | | | | | |
|---|---|---|---|---|---|
| | **Maximum Bending Stress (psi)** | | | | **Maximum Deflection (in)** |
| | **Bottom** | | **Top** | | |
| **Model** | **Stress** | **Δ (%)** | **Stress** | **Δ (%)** | **Deflection** | **Δ (%)** |
| Original | (1,155) | - | 1,225 | - | 0.008038 | - |
| Increased Mesh | (1,178) | 2.0% | 1,272 | 3.8% | 0.008049 | 0.1% |
| Decreased Mesh | (1,134) | -1.8% | 1,134 | -7.4% | 0.008030 | -0.1% |
| Simplified | (1,200) | 3.9% | 1,200 | -2.0% | 0.008000 | -0.5% |
| Note: Stress values in parentheses are compressive. | | | | | |

# 8.4   REPORTING/VERIFICATION CHECKLIST

## 8.4.1   Applicable to Reporting

☐   Is the summary material self-sufficient?

☐   Is the abstract in a text-based paragraph form for portability between computer platforms?

☐ Is the main report self-sufficient?

☐ Has the analysis been appropriately described to allow the analysis to be reconstructed at a later date (including references)?

☐ Have the appropriate technical references used in the approach been stated?

☐ Have all relevant assumptions pertaining to the analysis been included?

☐ Have the test method and conditions been described in sufficient detail to allow the testing to be repeated if necessary?

☐ Are the conclusions consistent with the results?

☐ Do the conclusions support the objectives?

☐ Do the tables and figures follow their reference in the text?

☐ Have any large tables provided as a matter of record been placed in the appendix?

☐ Has consideration been made for the readability of monochrome copies of color material?

☐ Does the reference material include the appropriate date or revision information?

☐ Has any obscure reference material been included in the appendix?

☐ Has any material included in the appendix been referenced in the body of the report?

## 8.4.2    <u>Applicable to Verification</u>

☐ Has the analysis been reviewed by qualified peers?

☐ Is the theoretical basis for the analytical approach fully understood?

☐ Are all of the assumptions (including underlying assumptions) consistent with the specific situation?

☐ Has all source material been verified against the references?

☐ Has verification testing of the analysis been conducted?

☐ Does testing reflect the expected scatter in results?

☐ Has the input data and mesh size of any finite-element models been verified?

☐   Has a simplified analysis been used to verify the results?

☐   Has the accuracy of any equations been verified against source material?

☐   Has the analysis been checked for mathematical errors?

☐   Has the effect of tolerances and property variations been assessed?

## 8.5   REFERENCES

[71]   Ulman, J. & Gould, J.; Technical Reporting; Holt, Rinehart and Winston, Inc. 1972

[72]   Ahrens, Jr., E.P.; "The Perils of Imprudent Writing"; Edward P. Ahrens, Jr. 1992

[73]   Singer, F.L.; Strength of Materials; Harper & Brothers, 1951

[74]   Materials Engineering, 1992 Materials Selector; Penton Publishing, Cleveland, OH; December 1991

# Appendix A - Abbreviations

| | |
|---|---|
| AC | Alternating Current |
| BGA | Ball-Grid Array |
| CAD | Computer Aided Design |
| CALCE | Computer-Aided Life Cycle Engineering (University of Md.) |
| CFD | Computational Fluid Dynamics |
| CG | Center of Gravity |
| CIC | Copper-Invar-Copper |
| CMC | Copper-Molybdenum-Copper |
| COB | Chip-on-Board |
| CSP | Chip-Scale Package |
| CTE | Coefficient of Thermal Expansion |
| DIP | Dual-Inline Package |
| DOF | Degree of Freedom |
| DC | Direct Current |
| IC | Integrated Circuit |
| IGES | Initial Graphics Exchange Specification |
| LCC | Leadless Chip-Carrier |
| PBL | Pressure Band Level |
| MCM | Multichip Module |
| PEM | Plastic-Encapsulated Microcircuit |
| PSD | Power Spectral Density |
| PSL | Pressure Spectrum Level |
| PTH | Plated-Through Hole |
| PWB | Printed Wiring Board |
| RMS | Root Mean Square |
| RSS | Root Sum of Squares |

| SMT | Surface-Mount Technology |
| S-N | Stress vs. Number of Cycles |
| SOIC | Small-outline Integrated Circuit |
| SPL | Sound Pressure Level |
| VOM | Volt-Ohm Meter |

# Appendix B - Unit Conversions

Familiarity for mechanical engineers is often with customary units, whereas electrical engineers use units that are based on the metric system. This leads to unusual units for thermal analysis where power is expressed in Watts, energy in Joules, temperature in degrees Celsius, dimensions in inches, and mass in pounds. Some useful unit conversions for various parameters are as follows:

**Acceleration**

$$1\ g = 9.807 \frac{meter}{sec^2}$$

$$1\ g = 32.2 \frac{ft}{sec^2}$$

$$1\ g = 386.1 \frac{in}{sec^2}$$

**Coefficient of Thermal Expansion**

$$\frac{1}{{}^\circ F} = \frac{1}{{}^\circ R} = \frac{1.8}{{}^\circ C} = \frac{1.8}{{}^\circ K}$$

**Convection Coefficient**

$$1 \frac{Btu}{hr\text{-}ft^2\text{-}{}^\circ F} = 3.663 x 10^{-3} \frac{Watt}{in^2\text{-}{}^\circ C}$$

$$1 \frac{Btu}{hr\text{-}ft^2\text{-}{}^\circ F} = 5.678 \frac{Watt}{meter^2\text{-}{}^\circ C}$$

**Density**

$$1 \frac{lbm}{in^3} = 27.68 \frac{g}{cm^3}$$

$$1 \frac{lbm}{in^3} = 2.768 x 10^4 \frac{kg}{meter^3}$$

$$1 \frac{lbm}{in^3} = 1,728 \frac{lbm}{ft^3}$$

**Energy**

$$1\ Btu = 252\ cal$$

$$1\ Btu = 1,055\ Joule$$

$$1\ Btu = 778.2\ ft\text{-}lbf$$

$$1\ kW\text{-}hr = 3,412\ Btu$$

**Force**

$$1\ lbf = 4.448\ Newton$$
$$1\ lbf = 32.2\frac{lbm\text{-}ft}{sec^2}$$
$$1\ lbf = 386.1\frac{lbm\text{-}in}{sec^2}$$

**Length**

$$1\ in = 0.0254\ meter$$
$$1\ in = 2.54\ cm$$
$$1\ ft = 30.48\ cm$$
$$1\ ft = 0.3048\ meter$$
$$1\ ft = 12\ in$$

**Mass**

$$1\ lbm = 0.454\ kg$$
$$1\ slug = 1\frac{lbf\text{-}sec^2}{ft} = 32.2\ lbm$$
$$1\frac{lbf\text{-}sec^2}{in} = 386.1\ lbm$$

**Power**

Note: Units for power also apply for heat transfer rate

$$1\frac{Btu}{hr} = 0.293\ Watts$$
$$1\ horsepower = 745.7\ Watts$$
$$1\ horsepower = 2,544\frac{Btu}{hr}$$

**Pressure**

Note: Units for pressure also apply to stress, and moduli of elasticity, plasticity, etc.

$$1\ psi = 6,895\frac{Newton}{meter^2} = 6,895\ Pascals$$
$$1\ psi = 2.036\ inches\ of\ mercury$$
$$1\ atm = 14.696\ psi$$
$$1\ atm = 1.013x10^5\ Pascals$$
$$1\ psi = 51.715\ mm\ of\ mercury = 51.715\ Torr$$

**Specific Heat**

$$1\frac{Btu}{lbm\text{-}^\circ F} = 4,187\frac{Joule}{kg\text{-}^\circ R}$$

$$1\frac{Btu}{lbm\text{-}^\circ F} = 1,899\frac{Joule}{lbm\text{-}^\circ R}$$

$$1\frac{Btu}{lbm\text{-}^\circ F} = 1\frac{cal}{gram\text{-}^\circ C}$$

**Spring Rate**

$$1\frac{lbf}{in} = 175.1\frac{Newton}{meter}$$

**Stefan-Boltzmann Constant**

$$0.1712x10^{-8}\frac{Btu}{hr\text{-}ft^2\text{-}^\circ R^4} = 5.669x10^{-8}\frac{Watt}{meter^2\text{-}^\circ K^4}$$

$$0.1712x10^{-8}\frac{Btu}{hr\text{-}ft^2\text{-}^\circ R^4} = 3.657x10^{-11}\frac{Watt}{in^2\text{-}^\circ K^4}$$

**Temperature**

Note: These conversions are valid for absolute and delta temperatures; conversions between actual Celsius and Fahrenheit must account for 0°C representing 32°F.

$$1^\circ K = 1.8^\circ R$$
$$1^\circ C(delta) = 1.8^\circ F(delta)$$

**Time**

$$1\ hr = 3,600\ sec$$
$$1\ hr = 60\ min$$

**Thermal Conductivity**

$$1\frac{Btu}{hr\text{-}ft\text{-}^\circ F} = 0.04396\frac{Watt}{in\text{-}^\circ C}$$

$$1\frac{Btu}{hr\text{-}ft\text{-}^\circ F} = 1.731\frac{Watt}{meter\text{-}^\circ C}$$

**Thermal Resistance**

$$1\frac{hr\text{-}^\circ F}{Btu} = 1.896\frac{^\circ C}{Watt}$$

**Velocity**

$$1\frac{ft}{sec} = 0.682\frac{mi}{hr}$$

$$1\frac{ft}{sec} = 1.097\frac{km}{hr}$$

$$1\frac{in}{sec} = 0.0254\frac{meter}{sec}$$

**Viscosity**

<u>Absolute Viscosity</u>

$$1\frac{lbm}{hr\text{-}ft} = 4.134x10^{-4}\frac{kg}{sec\text{-}meter}$$

$$1\frac{lbm}{hr\text{-}ft} = 4.134x10^{-3}\frac{gram}{sec\text{-}cm} = 4.134x10^{-3}\ Poise$$

<u>Kinematic Viscosity</u>

$$1\frac{ft^2}{hr} = 2.581x10^{-5}\frac{meter^2}{sec}$$

$$1\frac{ft^2}{hr} = 0.2581\frac{cm^2}{sec} = 0.2581\ Stokes$$

**Volume**

$$1ft^3 = 28.317\ liter$$

$$1ft^3 = 7.481\ gal$$

$$1\ liter = 61.02\ in^3$$

$$1\ gal = 231\ in^3$$

# Index

## A

acceleration due to gravity, 22–23, 65, 267, 268–269, 276

analysis planning, 13–21
  methods, 15, 44, 102
  tradeoffs, 15–21

assumptions, 21, 49, 51, 53, 58, 60, 69, 73–74, 75, 110, 122, 129, 181, 187, 192, 202, 226–227, 237, 238, 245, 248, 268, 269, 271–272, 330

## B

ball-grid array, 7–8, 94–95, 158, 213

boundary conditions
  mechanical, 105
  thermal, 46

## C

checklist
  analytical tool, 39–41
  energy-based methods, 277–282
  fire-resistance analysis, 277–282
  general analysis, 23
  humidity analysis, 277–282
  life analysis, 214–217
  mechanical analysis, 159–162
  other analysis, 277–282
  pressure drop analysis, 277–282
  pressure transducer rupture analysis, 277–282
  reporting, 342–343
  similarity analysis, 277–282

test data analysis, 323–324
thermal analysis, 96–98
verification, 343–344

chip-on-board, 10, 95, 159, 213–214

coefficient of thermal expansion, 130–131, 136, 138, 182–185, 186–192, 316–318, 320–321, 347

combustion processes, 220–221

convection, 62–69
  forced, 62–65, 223, 225–226, 228, 232
  free, 65–66

convection coefficient, 347

critical pressure, 236

cumulative damage, 169–170, 192

curvature, 126–128, 144, 156–158, 168–169

custom programs, 35–37

## D

damping, 111–112

deflection spectral density, 128

density, 136, 225, 230, 236, 249–250, 252–254, 257–258, 347

dew point, 244–245, 246–247

ductility, 182–185, 190

## E

elastic energy, 267, 268–269

electronic packaging, 1